# Fossils on the Seashore

# Fossils on the Seashore

## Beachcombing and Palaeontology

*Stephen K. Donovan*

LIVERPOOL UNIVERSITY PRESS

In memory of
George Sevastopulo (1941–2021) and Euan Clarkson (1937–2024) –
exemplary teachers and an inspiration to us all.

First published 2025 by
Liverpool University Press
4 Cambridge Street
Liverpool
L69 7ZU

British Library Cataloguing-in-Publication data
A British Library CIP record is available

ISBN 978-1-78046-109-0 (hardback)
ISBN 978-1-78046-122-9 (PDF)
ISBN 978-1-78046-123-6 (ePub)

Typeset by Carnegie Book Production, Lancaster
Printed in the Czech Republic via Akcent Media Limited

The manufacturer's authorised representative in the EU for product safety is:
Easy Access System Europe, Mustamäe tee 50, 10621 Tallinn, Estonia
https://easproject.com (gpsr.requests@easproject.com)

# Foreword

*Dr D.T.J. Littlewood*

*(Director of Science, The Natural History Museum, London)*

Armed with an enormous fossil oyster, I first met Steve Donovan almost four decades ago in his university office between lectures. What began as an enquiry soon developed into trips to the coast, with well-planned refreshments and celebratory stops, an introduction into deep time, past worlds, and ancient life forms and a long-lasting, cherished friendship of the highest calibre. Steve is a retired professional fossil sleuth and lifelong amateur, with enormous expertise, experience, skill and passion for field-based palaeontology.

Some of us have been lucky enough to be encouraged to look for nature's treasures from a young age, and some are even more fortunate to learn that curiosity and knowledge go hand in hand in revealing the richness of the many treasures around us. It's never too late to start. We each see things from different perspectives, and natural history specimens offer a perfect way to learn about the past and the present whilst also considering the future. *Fossils on the Seashore* is your companion to finding and collecting traces of the diversity of life entombed long ago, and offers you access to a lifetime of perspective. Finding delicate imprints of fragile soft-bodied invertebrates or stumbling over the enduring presence of shells and bones of ancient marine and terrestrial animals can all benefit from some friendly help. Welcome to the how, what and where of beachcombing nature's history and a window into the dynamic interplay between life and the geological forces that shape our world.

Standing in front of a fossil oyster bed, Steve brought to life an ancient ecosystem for me by interpreting the rocks and their entombed fauna, transporting us to prehistory and biodiversity long since gone. By asking questions and gathering evidence, we moved from curiosity to discovery and from information to knowledge – an everyday journey for a palaeontologist but a whole new dimension of natural history for me to appreciate. Almost hidden in plain sight, there's more natural history along the beach than you might imagine.

Steve's book is not just a collection of his insights, but a practical guide that will help you engage with the wonders of shoreline palaeontology. It will open new worlds of wonder as you collect and select the mineralised remnants of ancient life on sandy, muddy, pebbly or rock-strewn beaches. His conversational style and ability to make complex topics understandable make Steve the perfect guide for both seasoned collectors and newcomers.

As you embark on your own adventures, I hope you'll find the same joy and inspiration many amateurs and professionals have discovered alongside Steve. This book, with its practical advice, encouragement, tips on collecting, preserving and publishing fossils, and a friend to inspire and support you as you delve deeper into rocks and shorelines, will be as important as your hand lens and hammer.

So, as you prepare for your next trip to the beach or fossil hunting expedition, don't forget to add this book to your backpack. It will not only inspire you before your fossil forays but also inform and enthuse you during them. Consider it your trusted companion, guiding you through the fascinating world of palaeontology. Happy hunting!

# Acknowledgements

I retired on 2 October 2020, with my last book, *Hands-On Palaeontology*, in press; it was published in 2021. I was not fazed by the question of what to do in retirement. There were still a few papers to be finished from before 2020, but I had put many projects to sleep during my last 24 months of paid employment. In retirement, I seem to be alighting on just enough interesting geological ideas to keep my pen busy. I no longer travel to the Antilles for fieldwork, but the British Isles has ample geological distractions to keep me diverted. Like Orwell, I felt I might have one more book inside me, but, unlike George, I have survived to see its completion. But I am 70 now and the intention is to write research papers, not books, in my dotage. So, *Fossils on the Seashore* is the last of its kind. I hope it is fun to read.

As always, many thanks to all the colleagues, reviewers and editors of over 40 years, many of whom I am proud to call friend, who have supported, criticised, corrected and debated my ideas on palaeontology and geology. Anthony Kinahan at Dunedin Academic Press and his external assessor both liked my book proposal and said full steam ahead. Before I had finished, Dunedin was bought up by Liverpool University Press. I thus gained a new editor and thank Michael Ainsley for shepherding this volume to publication. My partner of more years than we care to admit to, Karen, provided support, encouragement and many, many cups of tea. The children, Hannah and Pelham, and my daughter-in-law Louisa, now all live in north-west England like me and provided much diversion, both wearing walking boots and in restaurants.

My modus operandi has changed in retirement. Unlike my last two books, *Fossils on the Seashore* was not written in a succession of cafés. Rather, I used the spare bedroom at Karen's apartment and the living room at my own, which also doubles as my reference library. I miss my many cafés in the Netherlands, but they are 500 km away on the other side of the North Sea. All were essential as I wrote my previous books, but I now dance to a different drum.

# Contents

## III Where to go and what to see

# Sources

With few notable exceptions, all the illustrations in this book are mine, either new or reproduced from my published papers. Full bibliographic references are provided in the captions of each and every image or diagram that has been previously published. My own illustrations, reproduced herein, appeared in the following research journals, magazines and chapters, which I am grateful to acknowledge for publishing my research papers and more general articles.

*Bulletin of the Geological Society of Norfolk* (Figs 2.1–2.6, 5.2–5.4, 7.4, 13.4, 20.4, 20.5)

*Bulletin of the Mizunami Fossil Museum* (Figs 7.2, 12.4)

*Caribbean Journal of Science* (Figs 7.1, 16.1, 16.3)

*Contributions to Tertiary & Quaternary Geology* (Figs 12.6, 14.5, 14.7)

*Deposits* (Fig. 10.8)

*Encyclopedia of Paleontology* (Fitzroy Dearborn, Chicago) (Fig. 14.3)

*Fossil Forum* (Figs 19.1, 19.5, 19.6)

*Geological Society of America Memoir* (Figs 16.5, 16.6)

*Geology Today* (Figs 1.3, 13.1, 17.1–17.4, 20.2, 20.3)

*Journal of Paleontology* (Figs 11.2, 14.1, 16.2)

*Journal of the Geological Society of Jamaica* (Fig. 14.4)

*Lethaia* (Fig. 11.4)

*Northumbrian Naturalist* (Figs 21.1–21.7)

*North West Geologist* (Figs 4.2–4.4, 7.3)

*Palaeontology* (Figs 11.3, 13.7)

*PalZ* (Fig. 14.2)

*Proceedings of the Geologists' Association* (Figs 1.1, 9.4, 11.1, 15.3)

*Proceedings of the Yorkshire Geological Society* (Fig. 5.1)

*Scottish Journal of Geology* (Figs 1.4, 22.1–22.3)

*Scripta Geologica* (Fig. 20.1)

*Swiss Journal of Palaeontology* (Fig. 9.7)

*The Processes of Fossilization* (Belhaven Press, London) (Fig. 16.4)

*Wight Studies: Proceedings of the Isle of Wight Natural History & Archaeological Society* (Figs 4.5–4.7, 15.4, 18.1–18.5)

Permission to reproduce the title page of the *Directory of British Fossiliferous Localities* (Fig. 1.2 herein) was granted by the editors of the Palaeontographical Society. Figures 10.4–10.7 are reproduced with the permission of the University of Chicago Press. Figures 8.3, 9.1–9.3, 9.5, 9.6, 10.1–10.3, 12.1, 12.3, 13.2, 13.3, 13.5, 13.6, 14.6, 15.1 and 15.2 are reproduced from textbooks, as recorded in the text, whose copyright has expired.

# I

# Introducing aspects of the beach

# CHAPTER 1

## Introduction: On the beach

[T]he very special world of the amateurs, an enormous band of unsung heroes and heroines, the real mainstay of natural history in Britain, each of whom has become expert in their own particular field in their spare time. (Bellamy, 2002, p. 113)

I miss my old friend Joe Collins (1927–2019) (Donovan & Mellish, 2020). We worked together on fossil crabs of the Antilles, mainly Jamaica. We had a fair division of labour. My jobs included collecting new specimens, to fit them into the geological succession, to take photographs and draw maps for any planned research paper, and so on. Joe was the crab systematist. I learnt most of what I know about fossil crabs from Joe, but he was the master. I would not have assumed to tell him his business.

Joe's life story was fascinating. He never spent a day studying in a university but had trained himself to be an internationally renowned expert on fossil crustaceans. He developed his interests in geology and palaeontology in the years following the Second World War. He was a founding member of an amateur group in South London, the Freelance Geological Society (later Association; FGA) (Donovan & Collins, 2016, 2020). And it was the escapades in the field by the FGA that really fired my imagination (Fig. 1.1).

**Figure 1.1** The FGA in the field. An expedition to the Cretaceous Gault Clay Formation at Reservoir Pit, Ford Place, Wrotham, Kent, c.1960 (after Donovan & Collins, 2016, fig. 7B). Members of the FGA are collecting systematically from the floor of the pit.

The south of England in the late 1940s and 1950s was a wonderland for the field palaeontologist. Joe's FGA was formed at the right time in the right place. There were still many working or recently discarded quarries. Transport was easy and affordable by rail and bus. Visits to quarries were rarely difficult – in the days before health and safety legislation became a junior synonym for no, get out, access was easily arranged. It was a world away from the mid-1970s, when I first picked up a geological hammer.

My favourite guide to where to collect was the marvellous *Directory of British Fossiliferous Localities* (Arkell *et al.*, 1954), published in the year of my own birth (Fig. 1.2). It was not surprising that 20+ years later some things had transformed. In 1954, there were still common quarries that were available to collectors; the many authors of *British Fossiliferous Localities* would have been familiar with them all. Then things changed. In 1975 or 1976, my friend Paul Vidler and I made a trip to Aylesbury, Buckinghamshire, to collect Jurassic ammonites. We found the site of Webster and Cannon's brickpit, guided by Arkell *et al.* (1954, p. 3) and the current Ordnance Survey 1:50,000 sheet, only to find it was completely infilled and levelled. Another time, a group of us 'bunked' into a disused Chalk pit near Cliffe, Kent (Arkell *et al.*, 1954, p. 57), only to find a notice on the way out that it was MoD

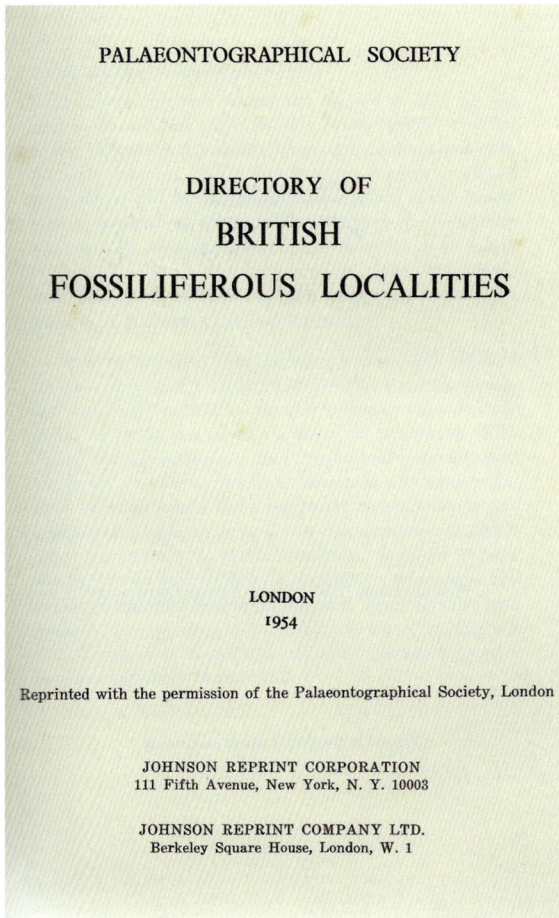

**Figure 1.2** Title page of the *Directory of British Fossiliferous Localities* (Arkell *et al.*, 1954). Author's collection.

PALAEONTOGRAPHICAL SOCIETY

DIRECTORY OF

**BRITISH**

**FOSSILIFEROUS LOCALITIES**

LONDON
1954

Reprinted with the permission of the Palaeontographical Society, London

JOHNSON REPRINT CORPORATION
111 Fifth Avenue, New York, N. Y. 10003

JOHNSON REPRINT COMPANY LTD.
Berkeley Square House, London, W. 1

property, and we were to keep out. Oops. We sauntered off with an air of innocence, with me happy with my first two tests of the heart urchin *Micraster* safely wrapped up in my bag. A happy day, indeed, but old quarries were not always welcoming sites for the collector even then.

The purpose of this long preamble is to emphasise how the search for fossils and fossil-iferous localities has changed as we look back from the perspective of the 2020s. In the 1940s and 1950s, collecting inland was relatively straightforward; there were numerous quarries, both working and disused. By the mid-1970s, these were disappearing, a trend that continues to the present day. Such is the state of quarrying in the UK that we import stone from overseas that was formerly sourced locally (Nield, 2014). Today, it is common for geologists and palaeontologists to access disused quarries, such as those preserved as Sites of Special Scientific Interest (SSSIs), rather than working pits that may be exposing new rocks and new information. Some road cuttings through fossiliferous rocks are accessible and safe if exploited sensibly. Similarly, the cuttings of disused railways may be productive if collected with care (such as Tracey *et al.*, 2002). There may be river exposures, too, but these are commonly valued more by the archaeologist than the geologist or palaeontologist (Maiklem, 2021). But, as the old song says, things ain't what they used to be. In the 2020s, the best place to practise field palaeontology in the UK is on the coast.

I admit that my early favourite localities were on the coast, at sites such as Copt Point, Folkestone, Kent (Cretaceous Gault Clay Formation; Hadland, 2018) and the Naze peninsula, Walton-on-the-Naze, Essex (Pliocene Red Crag Formation; Lee *et al.*, 2015). Care must be taken on cliffs, obviously (see Chapter 6), but common sense and caution both go a long way. Other beaches are a draw in themselves, with fossils reworked in cobbles and pebbles (Donovan, 2021, chapters 47, 48, 49). And there are recent shells on any beach. They are interesting in themselves, yet preserving features that inform our studies of classification, palaeoecology and taphonomy of fossils and trace fossils, such as encrusting organisms, borings and signs of predation. There is so much potential for you to make discoveries, and to expand your knowledge and experience. On the beach, we can truly use the present as the key to our fossil past.

If we take a suitably broad approach to palaeontology, any beach can be of relevance to our studies. If we wanted to be greedy, we might demand sites with *in situ* fossiliferous rocks, fossils in reworked beach clasts and interesting shells; Margate would fit this bill (Chapter 24; Fig. 1.3). But my own latest field campaign was on the east coast of the Isle of Mull in the Inner Hebrides (Chapter 30; Fig. 1.4). *In situ* fossiliferous rocks and limestone beach clasts are rare, and shells are of limited diversity and commonly broken. But this environment favoured a particular sort of study. The beaches, rich in igneous clasts, were a model for me to investigate the preservation of shells in a conglomeratic setting. I was delighted to find consistent patterns of breakage in some species of shells, whereas others were mainly complete. This is obviously an indication of selectivity in preservation. Thus, a beach without fossils had informed my ideas of preservation and the fossil record. I was delighted.

So, what is the purpose of *Fossils on the Seashore*? I want to encourage all palaeontologists to get into the field. I imagine that this book will appeal mainly to amateurs and students, but I hope that there is something for everyone. There are certain things I don't intend

**Figure 1.3** The coast near Margate, north Kent (after Donovan, 2020, fig. 1). The tide is creeping in on the beach below Cliftonville and looking west, back towards Margate. Chalk cliffs rise above the seawall to the left. Chalk clasts and modern shells can be collected mostly from between the highest point reached by dead seaweed on the beach and the sea (Chapter 17).

**Figure 1.4** Looking inland from Duart Bay, south-east Isle of Mull, at low tide (after Donovan, 2023, fig. 2). This is a pebbly, sandy beach with very rare, bored limestone clasts. Shells are not common and rarely well-preserved, but it has been informative for my interest in the taphonomy of shells in conglomeratic environments (Chapter 22).

to discuss in any detail, such as the geomorphology and formation of beaches; for this I recommend Pilkey *et al.* (2011). For discussion of British beaches at a non-technical level, you should try Carruthers & Dakkak (2020) and Holloway (2022).

Again, there are many books on the seashore, but none has a palaeontological focus. Old favourites with beachcombers include Soper (1972) and, more recently, Plass (2013). Others examine the shore as the margin of our terrestrial environment (Sprackland, 2012). Books on mudlarking are currently popular, and justly so, but their focus tends to be more archaeological than geological and more on the River Thames than the coast (Maiklem, 2020, 2021; Sandy & Stevens, 2021). Of all these, perhaps it is Soper that contains the most relevant information for the field palaeontologist, but it is now 50 years old. More applicable are books about pebbles on the beach (such as Ellis, 1954; Stocks & Lewin, 2019; Mitchell, 2021) and seashells (many titles, such as Barrett & Yonge, 1958; Street, 2019).

## References

Arkell, W.J. & 71 others. (1954) *Directory of British Fossiliferous Localities*. Palaeontographical Society, London.

Barrett, J.H. & Yonge, C.M. (1958) *Collins Pocket Guide to the Sea Shore*. Collins, London.

Bellamy, D. (2002) *Jolly Green Giant*. Century, London.

Carruthers, J. & Dakkak, N. (eds). (2020) *Sandscapes: Writing the British Seaside*. Palgrave Macmillan, Cham, Switzerland.

Donovan, S.K. (2020) Fossils explained 78: Never bored by borings. *Geology Today*, **36**: 232–235.

———. (2021) *Hands-On Palaeontology: A Practical Manual*. Dunedin Academic Press, Edinburgh.

———. (2023) Notes on *Aktuo-Paläontologie* of the rocky beaches of the eastern Isle of Mull, UK. *Scottish Journal of Geology*, **59**: 5 pp.

Donovan, S.K. & Collins, J.S.H. (2016) A brief history of the Freelance Geological Association (FGA), 1948–1967. *Proceedings of the Geologists' Association*, **127**: 90–100.

———. (2020) In the field with Joe: Early excursions of the Freelance Geological Society. *Geology Today*, **36**: 53–58.

Donovan, S.K. & Mellish, C.J.T. (2020) Mr Joseph Stephen Henry (Joe) Collins, 1927–2019. *Bulletin of the Mizunami Fossil Museum*, **46**: 103–114.

Ellis, C. (1954) *The Pebbles on the Beach*. Faber & Faber, London.

Hadland, P. (2018) *Fossils of Folkestone, Kent*. Siri Scientific Press, Manchester.

Holloway, S.McG. (2022) *The Beaches of Scotland*. Vertebrate Publishing, Sheffield.

Lee, J.R., Woods, M.A. & Moorlock, B.S.P. (2015) *British Regional Geology: East Anglia*. 5th ed. British Geological Survey, Nottingham.

Maiklem, L. (2020) *Mudlarking: Lost and Found on the River Thames*. Bloomsbury, London.

———. (2021) *A Field Guide to Larking*. Bloomsbury, London.

Mitchell, C. (2021) *The Pebble Spotter's Guide*. National Trust Books, London.

Nield, Ted. (2014) *Underlands: A Journey through Britain's Lost Landscape*. Granta, London.

Pilkey, O.H., Neal, W.J., Kelley, J.T. & Cooper, J.A.G. (2011) *The World's Beaches: A Global Guide to the Science of the Shoreline*. University of California Press, Berkeley.

Plass, M. (2013) *RSPB Handbook of the Seashore*. Bloomsbury, London.

Sandy, J. & Stevens, N. (2021) *Thames Mudlarking: Searching for London's Lost Treasure*. Shire, Oxford.

Soper, Tony. (1972) *The Shell Book of Beachcombing*. David & Charles, Newton Abbot.

Sprackland, J. (2012) *Strands: A Year of Discoveries on the Beach*. Cape, London.

Stocks, C. & Lewin, A. (2019) *The Book of Pebbles*. Thames & Hudson, London.

Street, P. (2019) *Shell Life on the Seashore*. Faber & Faber, London.

Tracey, S., Donovan, S.K., Clements, D., Jeffery, P. *et al.* (2002) Temporary exposures of the Eocene London Clay Formation at Highgate, north London: Rediscovery of a fossiliferous horizon 'lost' since the nineteenth century. *Proceedings of the Geologists' Association*, **113**: 319–331.

# CHAPTER 2

# Molluscs or pebbles?

## Introduction

I collect both modern molluscs shells and lithic clasts, but both to a purpose. My interests include palaeoecology, preservation and borings (see Chapter 5). I collect notable specimens which I then describe and publish. In order to publish my observations on these specimens, it is mandatory to deposit them in recognised museum collections (Donovan, 2021, p. 173). So, most specimens are only in my collection for a short while before being moved to a bigger and more public collection with a wider access for collectors and researchers.

To illustrate examples of what might be collected and why they might be of some general interest, I include two examples – one modern shells, one a lithic clast – that I found significant when I encountered them in the field. My observations and ideas were published, and the specimens are now available for examination in museum collections. You might follow the same path in publishing your own ideas (Donovan, 2017, 2021, pp. 165–175).

### Example 1 Molluscs

(Adapted after Donovan *et al.*, 2020.) Scientific discoveries may be made by plan or chance (Waller, 2004). For example, a specialist strolling along a beach may happen upon the remains of a specimen displaying features hitherto unknown to science and waiting description. Discoveries reliant on chance observations are informed encounters, not the fruit of a lengthy programme of research, yet worthy of study, nonetheless.

In this case the discoveries are not new species, but three specimens of common bivalve molluscs, each showing an unusual pattern of preservation and each allochthonous, washed up on the beach. This is a rare trio of bivalve specimens, each of which demonstrates a distinctive, perhaps unexpected pattern of pre- and post-mortem organism–organism interaction. If analogous specimens were disinterred from the rock record, they would stimulate significant questions concerning their rates of post-mortem burial.

Taxa were identified by reference to Barrett & Yonge (1977), Tebble (1976) and other relevant references. All specimens are registered in the Recent mollusc collections of the Naturalis Biodiversity Center, Leiden, the Netherlands (prefix RMNH.MOL).

*Locality*: Specimens were collected at the resort of Scheveningen in Den Haag (The Hague), from the beach on the Dutch North Sea coast (Fig. 2.1). The trend of the coast is north-east to south-west and the specimens were collected from the beach to the south-west, adjacent

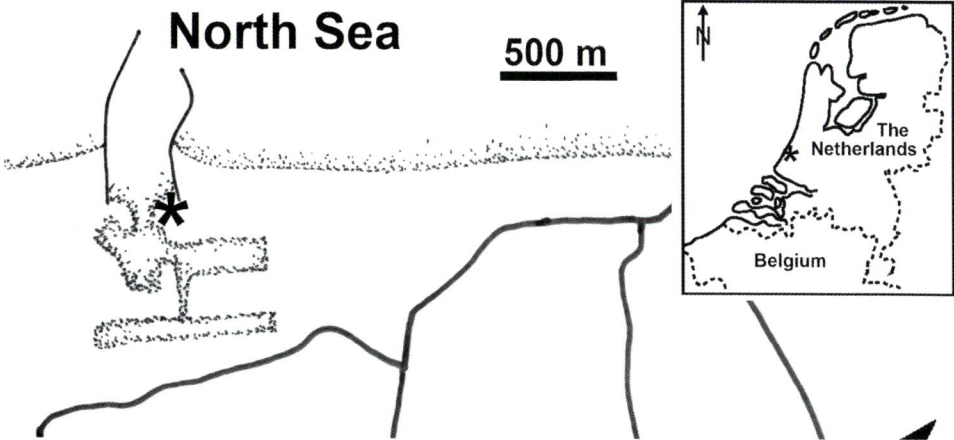

**Figure 2.1** (After Donovan *et al.*, 2020, fig. 1.) Outline map of the area around Scheveningen and its harbour (upper left) and part of the North Sea coast, Den Haag, the Netherlands. Key: stipple = coastline; * = shell locality; heavy lines = major roads. Inset shows position of main map (*) within the Netherlands.

to and within the harbour (52° 06′33.2″ N 04° 15′30.5″ E). Access to the beach was from Strandweg (= Beach Road). Specimens were collected both from the beach adjacent to the Noordelijk havenhoofd (= northern pier of the harbour) on its north-east side, which is part of the bathing beach, and from the small beach at the wall's south-east end within the harbour. They were presumably washed onshore from the shallow-deepening, sandy nearshore. Oyster valves are notably commoner within the protection of the harbour walls than outside them.

***Descriptions:*** *Mussel-oyster interaction* (Fig. 2.2): Both mussels and oysters are epifaunal. RMNH.MOL.340525 is comprised of both valves of an articulated(?) common mussel, *Mytilus edulis* Linné, which have been overgrown and stabilised by an oyster, *Ostrea edulis* Linné, of which only the cemented (left) valve is preserved. The mussel shell is preserved as a 'butterfly', that is, as an open shell with the valves gaping, but in close association (Ager, 1963, p. 84). After the adductor muscle scars rotted, the ligament was still intact, opening the valves and keeping them in close association. The shell was lying on the seafloor with the concave (internal) surfaces uppermost in order for it to be encrusted by the oyster, whose valve is cemented to the inner surfaces of both valves of the mussel. The umbo of the oyster is in the lower right of Figure 2.2A; that is, it attached to near the commissure of the mussel and on the far side away from the mussel's ligament. A raised 'knob' close to the umbo of the oyster is an overgrowth of the valve of a balanid barnacle (Fig. 2.2A, right, area of yellow discoloration).

*Cockle-barnacle interaction* (Fig. 2.3): Cockles are infaunal, balanids epifaunal. When collected, the valves of this edible cockle, *Cerastoderma edule* (Linné), were connected by the ligament which dried and broke before photography. The notable feature of this specimen is that it is the shell of a burrowing bivalve that is encrusted both externally and internally by the balanid barnacle, *Balanus crenatus* Brugière. The external surfaces of the

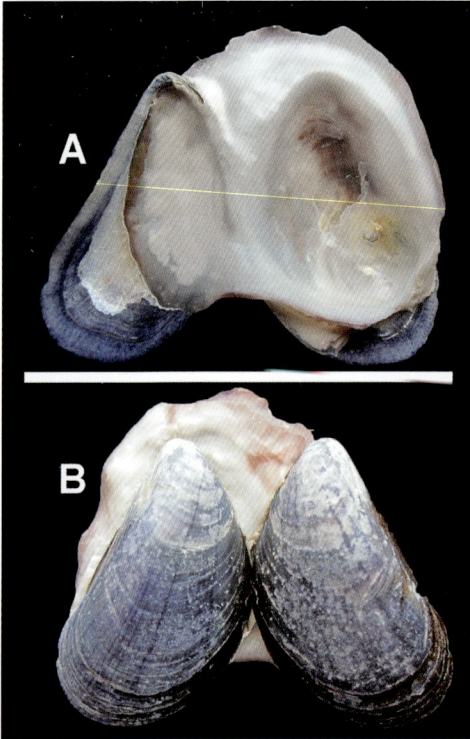

**Figure 2.2** (After Donovan *et al.*, 2020, fig. 2.) Modern dead shells, Scheveningen, North Sea coast, the Netherlands. RMNH.MOL.340525, the mussel shell, *Mytilus edulis* Linné, is preserved as a 'butterfly', and overgrown and stabilised by the oyster *Ostrea edulis* Linné. (**A**) Inner surface of mussel overgrown by the attached valve of the oyster (free valve lost). Umbo of oyster lower right; balanid overgrown by oyster close above the umbo. (**B**) Outer surfaces of mussel valves. The ligament of the mussel is not preserved, and the valves may have been rotated away from each other. Scale bar represents 50 mm.

**Figure 2.3** (After Donovan *et al.*, 2020, fig. 3.) Modern dead shells, Scheveningen, North Sea coast, the Netherlands. RMNH.MOL.340526, The valves of the cockle *Cerastoderma edule* (Linné) were connected by the ligament when collected, but it dried and broke subsequently. The cockle is encrusted both externally and internally by the balanid barnacle, *Balanus crenatus* Brugière. (**A₁**, **A₂**) Right valve, external (**A₁**) and internal surfaces (**A₂**). (**B₁**, **B₂**) Left valve, external (**B₁**) and internal surfaces (**B₂**). The large balanids on the external posterior surface (**B₁**) may have attached while the bivalve was in life position (but not necessarily alive), but in only a shallow burrow. Scale bar represents 50 mm.

valves preserve evidence of attachment scars (Fig. 2.3A₁, centre) and/or shells (Fig. 2.3A₁, lower centre; 2.3B₁, right = posteriorly) of cemented balanids. Internally (Fig. 2.3A₂, B₂), balanids were smaller, more plentiful and more widely distributed, but they subsequently all dropped off the smooth surface, although their attachment scars are still visible.

*Cockle-oyster interaction* (Fig. 2.4): RMNH.MOL.340527 preserves an interaction between an infaunal cockle and epifaunal oyster. The cockle, *C. edule*, is still articulated although the ligament has rotted away; that is, the valves are still in their relative positions as in life. This is because the oyster, *O. edulis* Linné, has overgrown the commissure and is cemented

**Figure 2.4** (After Donovan *et al.*, 2020, fig. 4.) Modern dead shells, Scheveningen, North Sea coast, the Netherlands. RMNH.MOL.340527, an articulated cockle, *Cerastoderma edule* (Linnaeus), with the valves still in their relative positions as in life. The oyster, *Ostrea edulis* Linnaeus, has overgrown the commissure and thus stabilised both valves of the cockle. Scale bar represents 50 mm.

to both valves of the cockle. The initial attachment appears to have been to the left valve (lower in Fig. 2.4) and included overgrowth of a few balanids. Both valves of the oyster are preserved; the right valve of the cockle is broken in the umbonal region (Schäfer, 1972, pp. 159–161, fig. 98).

**Discussion**: Bivalve shells 'are more likely to remain articulated under low-energy conditions and rapid burial' (Martin, 1999, p. 54). The specimens examined herein were not buried, and, in being found washed up on the beach, were obviously exposed to energetic conditions. However, this was most probably a rare storm event; the ambient conditions were most likely low energy, above or below wave base and otherwise protected by the walls of the harbour. Evidence from all specimens indicates a period of post-mortem residence on the seafloor, yet all were articulated when collected. It is this unusual pattern that is worthy of comment. Each demonstrates a mode of preservation which could potentially be recognised in the fossil record. Further, each specimen demonstrates (geologically) fast interactions between shelly organisms that might not have been anticipated. All are indicative of events that probably took only days or weeks.

The mussel (Fig. 2.2) lost its soft tissues post-mortem through rotting and/or scavenging, but the ligament had yet to break down, holding the valves in close association. A solitary balanid barnacle encrusted the inner surface of the left valve, as did an oyster. The mussel shell was a 'butterfly'. The ligament subsequently degraded, but not before the oyster had grown from the lower right in Figure 2.2A over the full width of this valve and onto the adjacent right valve. Encrustation by the oyster was thus faster than the time it took for the ligament to rot.

The cockle encrusted by balanids (Fig. 2.3) was, in life, infaunal. It must have been removed from the infaunal environment by either mechanical exhumation of a live or (more likely) dead shell, or by the live shell moving to the surface, probably shortly before death and possibly in response to adverse environmental conditions (Schäfer, 1972, pp. 158–159). At this time, and for the first time, the external surface of the shell would have been available to encrusting organisms, notably balanids. These are best preserved on the posterior of the left valve (Fig. 2.3B$_1$), which may have occurred when the cockle was still infaunal (Schäfer, 1972, fig. 69). Post-exposure attachment is indicated by basal scars in the centre of the right valve and remnants of a single balanid near the centre of the commissure (Fig. 2.3A$_1$). Later, subsequent to the loss of the soft tissues and the gaping of the shell, but while the ligament remained intact, the smooth interior surfaces of the shell were available for encrustation by a later balanid spatfall (Fig. 2.3A$_2$, B$_2$). Balanid shells and basal scars were numerous on the inner surfaces of both valves at the time of photography; all these barnacles were about the same size. In the 2+ years since collection, all the balanids on the internal surfaces have sloughed off, a post-mortem event that would probably leave little or no evidence if fossilised.

It is likely that the 'butterflied' RMNH.MOL.340526 was in its most stable position, concave-down, and the internal surfaces of the valves were settled by balanid larvae in this protected position (Schäfer, 1972, pp. 112–120). Single cockle valves, encrusted both internally and externally, are common fossil and Recent remains; what is most unusual is that the articulated shell shows analogous encrustations on all surfaces of the shell. That is, balanid infestation was fast, invading the dead shell after the loss of soft tissues, but before it could disarticulate. As such, it shows a similar pattern of infestation to some North Sea razor shells (Donovan *et al.*, 2014).

The infaunal cockle RMNH.MOL.340527 (Fig. 2.4) must have been at least partly exposed at the sediment surface in order to be encrusted by the epifaunal oyster (Schäfer, 1972, fig. 69). The initial overgrowth was by balanids. Presumably the cockle was dead or, at least, moribund for the oyster to successfully span and stabilise the commissure. The breakage of the umbo on the cockle's right valve is large and irregularly rounded with jagged edges. Breakage in this region is common in disarticulated valves (Schäfer, 1972, fig. 98) and was presumably due to mechanical impact post-mortem, perhaps during the storm that washed the specimen onto the beach; around the hole there is no sign of abrasion. The only other probability might be due to the impact of the beak of a herring gull (Schäfer, 1972, p. 414).

The specimens described above, taken together, represent a surprising trio of bivalves, although they can all be explained without excessive conjecture. This is further emphasised by them being found on a small part of a beach at Scheveningen at the same time. If any had been found as a fossil, they would have posed significant questions. That they were found together suggests that the post-mortem organism–organism interactions that they represent may be locally common. In part, this may be due to analogous specimens not being particularly rare but being ignored; the shells themselves are so common on Dutch beaches that nobody cares if anything abnormal has happened to them. So, why are analogous specimens at least rare, perhaps unknown in the fossil record? One aspect of their preservation is that they were all found on a beach, visible 'in the round' rather than as a two-dimensional section in a rock face. Further, they were presumably derived from shallow water, perhaps less than 10 m deep. Such palaeoenvironments are uncommon in the rock record; fossil beaches are notable, but difficult to beachcomb in the same manner as a modern analogue. If such notable interactions as these are limited to such shallow water environments at the present day, they were probably similarly so in the Cenozoic.

### Example 2 Lithic clasts

(Adapted from Donovan, 2023.) Why focus on a single specimen collected from a beach of many hundreds of cobbles and pebbles of mainly chalk and flint? Because this specimen demonstrates a combination of features, which, taken together, are worthy of close attention.

This specimen was immediately apparent on the beach because it included a fossil echinoid on the exposed side; it is likely that it would otherwise have been ignored. *Echinocorys* ex. gr. *scutata* (Leske) is the commonest of erratic echinoids derived from the Chalk and known from the north Norfolk coast (Donovan & Lewis, 2011), but they are still rare fossils in this setting. Thus, it was immediately intriguing to find this specimen, even before it was turned over. The specimen was collected on the beach between Overstrand and Cromer, north Norfolk (Chapter 20; Fig. 20.1).

**Description:** This clast (Figs 2.5, 2.6) is about 119 × 113 × 77 mm. The end of the clast (bottom, Fig. 2.5) at the opposite end to the echinoid (upper left, Fig. 2.5) was broken off in the field because it preserved no additional features of interest not already seen on Natuurhistorisch Museum Maastricht, the Netherlands, NHMM 2020 010. The bulk of the specimen was thus reduced to a manageable size.

In Figure 2.5, the cobble exposes the more or less complete test of *E.* ex. gr. *scutata*. Damage to the test is entirely mechanical in this orientation, including fractures in the test of the echinoid. The chalk and the echinoid on this side of the cobble show no sign of boring.

Rotation of this specimen through 90⁰ (as viewed from the upper end in Fig. 2.5) exposes a very different aspect. The surface examined in Figure 2.5 is truncated by a highly bored surface, dominated by the clionaid sponge trace *Entobia* isp. The borings infest the chalk, the test of *E.* ex. gr. *scutata* and, where exposed, the infill of the same. The sponge borings also preserve rare post-mortem nestling invertebrates, a small spirorbid and clustered serpulids. The arbitrarily determined upper/lower surface divide is indicated by the absence/presence of *Entobia*.

Further rotation through 90⁰ exposes the entire surface of the reverse side to that shown in Figure 2.5 (Fig. 2.6). This is dominated by *Entobia* isp. Indeed, *Entobia* isp. is only absent where the surface has been abraded (left of centreline) or where it is bored to a deeper level by two incomplete *Gastrochaenolites* isp. (top centre). The narrower of these two borings (left, apparent in this orientation as a dark slot) is encrusted by multiple, spiral spirorbid worm tubes.

**Discussion:** NHMM 2020 010 is a 'busy' cobble which has undergone a range of events since exhumation (Figs 2.5, 2.6). These may be deduced from the many preserved features. It was broken away from an exposure of chalk, either nearshore and submerged or onshore, perhaps in a cliff. It was then shaped – rounded without becoming spherical, probably under partial control of the original bedding – by being rolled around by wave action, that is, by corrasion (corrosion+abrasion). This exposed much of the test of *Echinocorys* apart from the oral surface and part of the apex, which are both still concealed by rock. The cracks in the test (Fig. 2.5) are probably recent, otherwise they would have led to the spalling off of some parts.

The clast came to rest with the surface in Figure 2.5 buried in modern seafloor sediment, thus protecting it from borers and encrusters. In stark contrast, the reverse surface (Fig. 2.6) was exposed to any and all infesting organisms. The demarcation between these surfaces is sharp.

The exposed chalk and echinoid substrate proved to be ideal for boring organisms, as would be expected (Donovan & Lewis, 2011). As is quite common, the surface was invaded by just one dominant infesting, boring organism, in this example the clionaid sponge

**Figure 2.5** (After Donovan, 2023, fig. 2.) NHMM 2020 010, lateral view of cobble showing test of an irregular echinoid, *Echinocorys* ex. gr. *scutata* (Leske) (upper right). This is a lateral view of the echinoid; its apex is towards the top right. Note fractured surface of the test. This clast was broken, towards the bottom in this orientation, by the author. Scale in cm and mm.

boring *Entobia* isp. At that time, the surface of the chalk would have been perforated by small sponge apertures only (similar to Donovan, 2021, fig. 48.5C). The current labyrinthine geometry of the surface (Fig. 2.6) is a result of later corrasion of the cobble, aided by the weakening of the surface by intensive boring. Presumably prior to this later corrasion, two boring bivalves perforated the bored surface (= *Gastrochaenolites* isp.).

When exposed on the seafloor, the test of *Echinocorys* in the cobble would have been largely complete and well preserved. However, the side of the test exposed on the seafloor was an equally good substrate for boring sponges as was the chalk. Most of the test on this side is destroyed, but its outline is subtly apparent because its chalk infill, now exposed and bored, is slightly darker than the surrounding sedimentary rock (Fig. 2.6).

After the death of the boring organisms and some subsequent corrasion, the chalk substrate became available to 'worms' and their calcareous cemented tubes. These are small and sparse. The most common encrusters were spirorbids, which are particularly prominent

**Figure 2.6** (After Donovan, 2023, fig. 3.) NHMM 2020 010, lateral view of clast showing modern borings; bored *E.* ex. gr. *scutata* bottom right. The borings are mainly *Entobia* isp. Two incomplete *Gastrochaenolites* isp. occur in a depression, top centre. The top of this view corresponds to the bottom in Figure 2.5. Scale in cm and mm.

in one of the *Gastrochaenolites* (only seen as an oblique 'slot' left of centre in the upper part of Fig. 2.6). These were most likely a late-stage infestation. This surface was in the sediment on the beach, which would have killed any surviving worms by smothering.

The bored side of this clast preserves only a small diversity of traces, but in a definite succession. The surface was initially completely infested by clionaid sponges (= *Entobia* isp.). These borings are widespread on this surface, but shallow; corrasion of the surface has exposed the interconnected chambers just beneath the original bored surface. This surface was subsequently perforated by two boring bivalves (= *Gastrochaenolites* isp.). These extended deeper into the chalk than the surface infestation of clionaids. These now incomplete borings are in a depression that is likely corrasional, in part, where chalk has spalled away. The walls of the *Gastrochaenolites* were partially infested by *Entobia* in this depression after the death of the producing bivalves (Fig. 2.6). Most particularly, one of the *Gastrochaenolites* is perforated by rare apertures of *Entobia* (in stark contrast, the other has a dark lining encrusted by spirorbids). Thus, *Gastrochaeonolites* bored *Entobia* and the pristine chalk beneath, but, and subsequently, the reverse was also true.

Does NHMM 2020 010 have any greater importance than the many other chalk clasts on the north coast of Norfolk? Yes, because of its mixed suite of palaeontological and neoichnological features preserved in close association. These have permitted the history of this clast to be determined in some detail. It is the combination of features that makes this clast of broad interest.

## References

Ager, D.V. (1963) *Principles of Paleoecology.* McGraw-Hill, New York.

Barrett, J. & Yonge, C.M. (1977) (first published 1958) *Collins Pocket Guide to the Sea Shore.* Collins, London.

Donovan, S.K. (2017) *Writing for Earth Scientists: 52 Lessons in Academic Publishing.* Wiley-Blackwell, Chichester.

———. (2021) *Hands-On Palaeontology: A Practical Manual.* Dunedin Academic Press, Edinburgh.

———. (2023) Three views: Complex post-exhumation history of a Chalk cobble, north Norfolk. *Bulletin of the Geological Society of Norfolk*, **73**: 85–93.

Donovan, S.K., Cotton, L., Ende, C. van den, Scognamiglio, G. & Zittersteijn, M. (2014) Taphonomic significance of a dense infestation of *Ensis americanus* (Binney) by *Balanus crenatus* Brugière, North Sea. *Palaios*, **28** (for 2013): 837–838.

Donovan, S.K., Hoeksema, B.W., Fransen, C.H.J.M., Vonk, R. & Adema, J.P.H.M. (2020) Unusual preservation of North Sea shells: Scheveningen, North Sea coast, the Netherlands. *Bulletin of the Geological Society of Norfolk*, **70**: 55–65.

Donovan, S.K. & Lewis, D.N. (2011) Strange taphonomy: Late Cretaceous *Echinocorys* (Echinoidea) as a hard substrate in a modern shallow marine environment. *Swiss Journal of Palaeontology*, **130**: 43–51.

Martin, R.E. (1999) *Taphonomy: A Process Approach.* Cambridge University Press, Cambridge.

Schäfer, W. (1972) *Ecology and Palaeoecology of Marine Environments.* University of Chicago Press, Chicago.

Tebble, N. (1976) *British Bivalve Seashells.* 2nd ed. HMSO, Edinburgh.

Waller, J. (2004) *Fabulous Science: Fact and Fiction in the History of Scientific Discovery.* Oxford University Press, Oxford.

# CHAPTER 3

# Sandy beaches

## Introduction

After two chapters we are finally on the beach. Beach is a word that I use loosely to include a particularly wide range of shoreline environments. It is not too much of an exaggeration to say that any given beach is unique, different from every other beach, while recognising that there are many gradations and similarities. I might define a beach as 'a deposit of sediment ranging in [grain] size from [mud to] sand to boulders, formed by waves along a coastline' (Pilkey *et al.*, 2011, p. 27). I identify three broad categories of beach – sandy, muddy and beaches with common lithoclasts (pebbles, cobbles and boulders).

### Grain size: what is sand?

(Adapted from Donovan, 2021, pp. 24–25.) Siliciclastic sedimentary rocks are classified on the basis of grain size (Table 3.1). Pebbles, cobbles and/or boulders on a beach are indicative of a particularly high-energy setting in which clasts were transported and became rounded by abrasion. Such grains go to make rocks called conglomerates or, if clasts are angular, breccias (Donovan, 2023).

## Table 3.1 A grain-size classification of clastic sedimentary rocks, compiled from various sources

| Rock type | Grain size |
|---|---|
| Boulder conglomerate/breccia | > 256 mm |
| Cobble conglomerate/breccia | 64–256 mm |
| Pebble conglomerate/breccia | 4–64 mm |
| Gritstone | 2–4 mm |
| Sandstone: coarse-grained | ½ to 2 mm |
| Sandstone: medium-grained | ¼ to ½ mm |
| Sandstone: fine-grained | $\frac{1}{16}$ to ¼ mm |
| Siltstone | $\frac{1}{256}$ to $\frac{1}{16}$ mm |
| Mudrocks (shale, mudstone) | $< \frac{1}{256}$ mm |

(After Donovan, 2021, table 7.1). Most grain sizes can be determined in the field with the naked eye and hand lens. Siltstones and mudrocks are differentiated in the field by rubbing them, gently, over your teeth; mudrocks will be smooth, but siltstones will grate (but not too much) due to their fractionally larger grains, some of which may be apparent with a hand lens. Shales are fissile whereas mudstones are more massive. A conglomerate has more or less rounded grains; a breccia is conglomerate-like with angular grains.

Conglomerates are rarely fossiliferous, certainly more rarely than sandstones, siltstones and mudrocks. This is explained by grain size and energy/environment of deposition. A current flow capable of carrying pebbles and larger grains is equally likely to ground shells down to small, unidentifiable fragments through frequent collisions. A coarse-grained sandstone is still indicative of high-energy deposition, but some shells, at least, will be much larger than individual sand grains, and, in consequence, relatively more durable. A beach with common pebbles and cobbles will probably have only a relatively sparse shelly component because they have all been ground down by movement of pebbles and cobbles. So, if you intend to collect modern shells, sandy and muddy beaches are likely to provide the best and most numerous specimens (Donovan, 2023).

## Provenance

Where does the sediment of a beach come from? Some grains may derive more or less *in situ*. The Port Royal Cays, offshore Kingston Harbour, Jamaica, have white carbonate sand beaches. Any handful of sediment is rich in identifiable bioclasts; for example, it may be full of spines of regular echinoids but ground down to more or less rounded fragments and, thus, not a threat to exposed skin. These grains must have been derived locally, just offshore, as the elevation of the cays is only a metre or so and there is no source of land-derived sediment, being surrounded by deeper water. Transport of sand grains must have been minimal, although reworking by wave action was sufficient to corrade most calcareous skeletal bioclasts.

More commonly, and given a more extensive associated area of land, beach grains may derive sediment from inland, from offshore, laterally or, probably, a combination of all three. Consider one of the beaches discussed herein, that between Overstrand and Cromer on the north Norfolk coast (Chapter 20; Donovan, 2021, pp. 195–199; Fig. 3.1 herein). The littoral or longshore drift of this coast – that is, the seaborne transport of sediment – is towards the west (Davies & Waterhouse, 2023). Thus, the lateral movement of grains is from the east. The beach also accumulates sediment from the cliffs at their inland margin, part of the Cromer Ridge, through large and small landslides which provide a mixture of Neogene siliciclastic grains and Chalk, including flints. There must also be movement onshore of similar rocks from the shallow sea, certainly by wave action, but particularly by winter storms. Storm-derived clasts may be large (Fig. 3.2).

The resultant beach is predominantly sand, but with many chalk and, particularly, flint pebbles and cobbles, with rare erratics derived by fluvial action prior to the Ice Ages (Donovan, 2010, and references therein). Obviously, the smaller and lighter that a grain may be, the easier it is to transport. Sand-sized grains will be more susceptible to longshore drift than a cobble of flint or chalk. So, with the migration of sand and smaller grains to the west on each tide, the nature of the beach is changed subtly with each tide or storm.

One final comment about sorting. The Chalk cliffs of this section contain a few flint horizons but are dominantly chalk. In contrast, the cobbles on the beach are dominantly flint with relatively few chalk clasts. The reason why is obvious. Flint is much harder than chalk, so it persists while the softer chalk clasts are ground away.

To give another Jamaican example well known to me, a good demonstration of the problems of determining provenance is provided by the black sands of Farquhar's Beach

**Figure 3.1** A small part of the beach between Overstrand and Cromer, north Norfolk, March 2024. Note the sandy beach with numerous lithoclasts (mainly flint) and thick, dark brown deposits of the sea mat or hornwrack (= bryozoan) *Flustra* sp. (see Chapter 9). Length of cane (left) about 0.92 m.

**Figure 3.2** A large, bored clast (= boulder) of chalk from the beach between Overstrand to Cromer, about 200 × 148 × 74 mm (see also Donovan, 2024, fig. 2). Such a rock, both rounded and impregnated by marine borings (small, slot-shaped cross-sections of *Caulostrepsis* isp.; large round holes, partial *Gastrochaenolites* isp.) must have been derived from offshore by a storm. Collection of NHMM 2024 008. Scale in mm.

(Donovan *et al.*, 1989) on the central south coast (Donovan *et al.*, 2010, fig. 1). The sand at Farquhar's Beach is composed primarily of titano-magnetite and titano-haematite crystals with lesser amounts of feldspar, quartz and calcite (McFarlane, 1977). Major element components in the magnetic portion of the black sands are $Fe_2O_3$ (85.7 per cent to 71.9 per cent), FeO (14.2 per cent to 2.2 per cent) and $TiO_2$ (16.3 per cent to 8.9 per cent) (Chubb, 1960).

Because the black sands are concentrated at the mouth of the Rio Minho and westward along the south coast from Farquhar's Beach in the parish of Clarendon to Long Acre Point in the parish of St Elizabeth, Chubb (1960) suggested that the source of the sands was most likely the Cretaceous inliers in the interior of central Jamaica. Erosion and transportation by rivers flowing southwards, accompanied by westerly longshore currents, accounts for the dispersion of sand along the coast. Wood (1976, p. 27) noted that sediment samples from Milk River Bay are almost totally composed of clastic material, although the cliffs that border Farquhar's Beach are dominantly limestones.

McFarlane (1977) proposed that the source of the black sands was an otherwise unknown igneous and/or metamorphic outcrop located south of the present coastline and not far from the present location of sand deposits. McFarlane suggested that these rocks had been exposed during the low sea level stands of the Pleistocene and erosion proceeded mainly by physical, rather than chemical, weathering. However, this theory must be considered, at best, speculative.

## Beachrock

(Adapted from Donovan *et al.*, 1993.) Beachrock is a notable feature of certain modern beaches. It is produced predominantly in tropical, carbonate-rich, intertidal environments by penecontemporaneous cementation in the zone between high and low tides (Scoffin, 1987). Beachrock is commonly composed of mainly carbonate grains with a carbonate cement, either aragonite and/or high-magnesium calcite (Stoddart & Cann, 1965), but non-carbonate beaches can similarly be lithified (Scoffin, 1987; Donovan *et al.*, 1993). The presence of man-made artefacts indicates that lithification of at least some beachrocks was very recent (certainly 20th century). For example, Scoffin (1970) recorded a conglomeratic beachrock in the Bahamas in which the clasts comprised shells of the large marine gastropod *Strombus*, bottles, fragments of coral and boulders of limestone with carbonate sand.

## References

Chubb, L.J. (1960) The black sands of Jamaica. Unpublished report, Geological Survey Department, Jamaica.

Davies, J.A. & Waterhouse, D.M.G. (2023) *Exploring Norfolk's Deep History Coast*. History Press, Cheltenham.

Donovan, S.K. (2010) A Derbyshire screwstone (Mississippian) from the beach at Overstrand, Norfolk, eastern England. *Scripta Geologica, Special Issue*, **7**: 43–52.

———. (2021) *Hands-On Palaeontology: A Practical Manual*. Dunedin Academic Press, Edinburgh.

———. (2023) Notes on *Aktuo-Paläontologie* of the rocky beaches of the eastern Isle of Mull, UK. *Scottish Journal of Geology*, **59**: 5 pp.

———. (2024) Life offshore: Evidence from a bored Chalk boulder. *Mercian Geologist*, **21**: 42–45.

Donovan, S.K., Blissett, D.J. & Jackson, T.A. (2010) Reworked fossils, ichnology and palaeoecology: An example from the Neogene of Jamaica. *Lethaia*, **43**: 441–444.

Donovan, S.K., Jackson, T.A. & Littlewood, D.T.J. (1989) Report of a field meeting to the Round Hill region of southern Clarendon, 9 April 1988. *Journal of the Geological Society of Jamaica*, **25** (for 1988): 44–47.

Donovan, S.K., Williams, R.A. & Rocke, J.A. (1993) Preservation of a clypeasteroid echinoid in Holocene beachrock, Jamaica. *Caribbean Journal of Science*, **29**: 264–267.

McFarlane, N. (1977) The non-carbonate Pleistocene sand deposits of the south central coast of Jamaica. Abstract, 10th INQUA Conference, University of Birmingham, England, August.

Pilkey, O.H., Neal, W.J., Kelley, J.T. & Cooper, J.A.G. (2011) *The World's Beaches: A Global Guide to the Science of the Shoreline*. University of California Press, Berkeley.

Scoffin, T.P. (1970) A conglomeratic beachrock in Bimini, Bahamas. *Journal of Sedimentary Petrology*, **40**: 756–759.

———. (1987) *An Introduction to Carbonate Sediments and Rocks*. Blackie, Glasgow.

Stoddart, D.R. & Cann, J.R. (1965) Nature and origin of beach rock. *Journal of Sedimentary Petrology*, **35**: 243–247.

Wood, P.A. (1976) Beaches of accretion and progradation in Jamaica. *Journal of the Geological Society of Jamaica*, **15**: 24–31.

# CHAPTER 4

## Pebbles, cobbles and oysters on the beach

### Introduction

A sandy beach with loose pebbles is a favourite place of mine for collecting. It will have a potential wealth of possibilities; diverse shells washed onshore and cobbles of sedimentary rocks that may include fossils and can be bored or encrusted by recent organisms. For the reader of this book, it holds the prospect of excellent collecting. Indeed, the sites discussed in Chapters 17 to 22 all fit into this category, more or less.

But why include oysters with lithic clasts? It is because of their size, structure and resilience. Oyster valves are commonly large and robust, unlike most bivalves. To the collector they are more akin to clasts of sedimentary rock than to other, smaller mollusc valves. Scallop valves can be analogous but are less common than oysters.

The two sites discussed below are considered in more detail in Donovan (2021). They reappear here because of their potential for interesting lithic and oyster substrates, and because they are favourite localities of mine.

#### Example 1 Cleveleys, Irish Sea Coast

(Adapted from Donovan, 2020.) A fossil boring may be well preserved, but it is rarely so easy to collect as a bored clast on a beach. A modern boring may not be complete, due to degradation of clasts by corrasion, but it is freely available for study through all possible angles, may preserve the producing organism or post-boring inhabitants (squatters) (for example, Donovan 2017, figs 2A, B and 3A–H, K, respectively), can be present in large numbers and can be moulded by various means (liquid latex rubber, plasticine or many alternatives) (Donovan 2017, fig. 3I, J).

The best place to look for modern borings is on the beach and the best substrates are various. Herein, the Recent borings of a beach site in north-west England are examined with the eye of a palaeontologist in order to glean ichnosystematic and ecological data that may be of relevance to the fossil record.

*Locality, horizon, material and methods:* Cleveleys, north of Blackpool, in Fylde, north-west Lancashire, has a coastline trending north–south and fronting the Irish Sea. The south end of the present study [about NGR SD 312 430] was easily reached by a short walk from the Cleveleys stop of the Blackpool to Fleetwood tram (Donovan, 2021, pp. 200–203). Trending north towards Rossall Beach [NGR SD 313 444], the beach, particularly in its upper parts adjacent to the sea wall, has an abundance of cobbles and pebbles of diverse lithologies. At least some of these are erratics reworked out of glacial deposits and derived from the Lake District (Ellis, 1968, p. 144) (Fig. 4.1).

The erratics considered herein are Lower Carboniferous (Mississippian) limestones, which outcrops on all sides of the Lake District except to the west (Murphy, 2015). The rare identifiable fossils in limestone clasts from Cleveleys are mainly Mississippian colonial corals such as *Syringopora* sp. and lithostrotionids (Donovan, 2021, fig. 48.4).

***Ichnology:*** *Caulostrepsis* isp. is mainly produced by annelids of the family Spionidae (such as *Polydora* spp., common around the coast of the British Isles), but also other polychaete worms (Bromley 2004, p. 460). *Caulostrepsis* isp. cf. *C. taeniola* Clarke (Figs 4.2A–C, E, F, 4.3B, C, E) was identified on 13 limestone clasts. It is gregarious, most commonly apparent as small, slot-like boreholes, elliptical to figure-of-eight shaped. Close examination shows that the shaft is divided by a central, longitudinal vane. Where the boring is exposed in longitudinal section, it is an elongate U-shape (Fig. 4.2E). Almost all specimens are only seen in transverse section (this is probably not the true aperture due to corrasion), which is inadequate for confident identification to ichnospecies.

*Caulostrepsis* penetrates some *Entobia* (Fig. 4.3B), but not others (Figs 4.2D, 4.3C, E). This suggests that many *Caulostrepsis* – *Entobia* associations were coeval, and the producing organisms were living in close association. Where *Caulostrepsis* penetrates *Entobia*, presumably the sponge producing the *Entobia* was dead at the time of infestation by the former.

**Figure 4.1** On the beach at Cleveleys, looking out to sea. The large limestone blocks are shore defences. Note the many rounded pebbles and cobbles on the beach.

**Figure 4.2** (After Donovan, 2020, fig. 2.) Mainly *Caulostrepsis* isp. cf. *C. taeniola* Clarke.
(**A**) RGM.1332343, cobble with common *C. taeniola* seen mainly in cross-section, borings appearing
slot-shaped or a figure-of-eight. (**B**) RGM.1332346, cobble densely infested, but with obvious areas
free of borings. (**C, D**) RGM.1332347, two sides of a cobble, one with sparse *Caulostrepsis* (**C**) and the
other with mainly mature *Entobia* isp. (**E**) RGM.1332348, several borings are open in longitudinal view,
showing parallel canals separated by a central vane and, towards 2 o'clock, the U-shaped termination.
(**F**) RGM.1332351, borings in the limestone, but not in the calcite veins, such as top centre to 4 o'clock.
Specimens uncoated. Scale bars represent 10 mm.

**Figure 4.3** (After Donovan, 2020, fig. 3.) *Caulostrepsis* isp. cf. *C. taeniola* Clarke (**A–C, E**) and *Entobia* isp. (**C–F**). (**A, B**) RGM.1332353, two sides of a cobble, sparsely (**A**) and intensely bored (**B**). (**C, E**) RGM.1332356 and 1332359, respectively, numerous *Caulostrepsis* that do not bore into mature *Entobia* isp. (**D**) RGM.1332357, well-exposed, colonial boring system, the outermost few mm of limestone having been lost through corrasion; more mature borings towards top. (**F**) RGM.1332360, mainly apertures of sponge colony. Specimens uncoated. Scale bars represent 10 mm.

*Entobia* isp. is produced by sponges, mainly members of the family Clionaidae (Bromley, 2004, p. 459), and was identified in 16 limestone clasts; it is numerous in most of these specimens. The appearance of *Entobia* isp. is highly variable, partly due to differences in the style of preservation and also because of differences in maturity of sponge colonies which can radically change in morphology with age. Surfaces that are essentially uncorraded since infestation are perforated by numerous apertures (Fig. 4.3F). Corrasion only needs to remove the outer few mm of a clast to reveal the complexities of the 3D structure of the sponge borings (examples include Figs 4.2D, 4.3C–E). Most specimens appear to be mature;

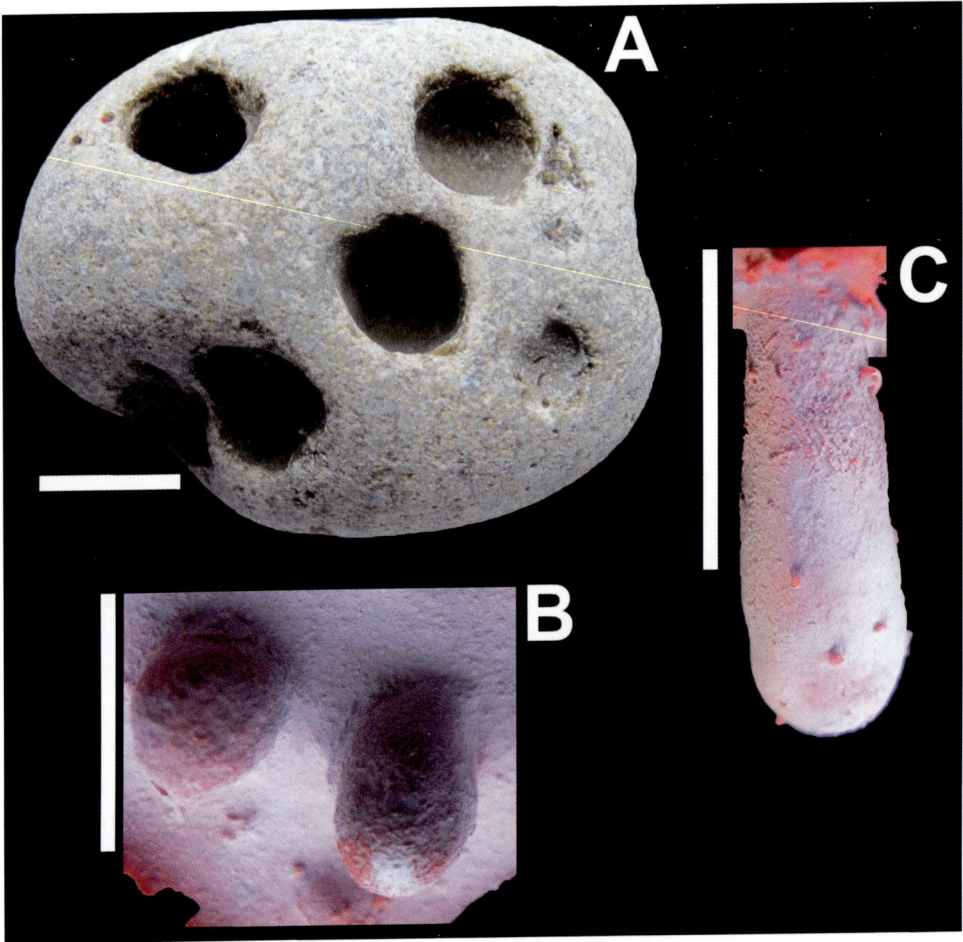

**Figure 4.4** (After Donovan, 2020, fig. 5.) *Gastrochaenolites turbinatus* Kelly & Bromley. (**A, B**) RGM.1332369. (**A**) Surface with shallow remnants of bivalve borings. (**B**) Latex moulds of two of the more complete borings on this surface. (**C**) RGM.1332370, latex mould of the most complete boring of *G. turbinatus* from this locality. Latex moulds coated with ammonium chloride. Scale bars represent 10 mm.

chambers having grown to be in lateral contact. In maturity, some chambers still appear to be spherical (Fig. 4.2D); they may have coalesced to produce a branching, tubular structure; and some have reached the stage of a 3D meshwork. A rare exception is seen in Figure 4.3D, in which chambers are still linked by slender lateral canals, with a more mature colony towards the top of the page.

*Caulostrepsis* and *Entobia* are the two most common ichnotaxa in the limestone clasts at Cleveleys. The developmental differences between separate specimens of *Entobia* are difficult to assign to ichnospecific rank.

*Gastrochaenolites* ispp. is mainly the spoor of boring bivalves (Bromley 2004, p. 462). *Gastrochaenolites turbinatus* is a rare boring at this site (Fig. 4.4). Whereas *Caulostrepsis* and

*Entobia* were both common, and only the best-preserved specimens were collected, only two clasts were found with *G. turbinatus*. RGM.1332369 preserves a suite of borings that are all very incomplete; only RGM.1332370 is sufficiently complete that a confident ichnospecific assignment is possible.

***Discussion***: The most important insight into the significance of these borings is provided by comparison with the conclusions of Donovan *et al.* (2019). The three ichnogenera documented above, namely *Caulostrepsis*, *Entobia* and *Gastrochaenolites*, are a common association today on the coasts of the southern and western North Sea, the English Channel and Irish Sea. These three ichnogenera have the advantage that they can be easily identified by the novice ichnologist. They form part of the suite of allochthonous ichnotaxa forming the *Trypanites* ichnofacies, indicative of a bioerosional, domichnial assemblage in a shallow water, rocky setting (Santos *et al.*, 2011) and, in this example, transported onshore by wave and storm action.

## Example 2 Queen Victoria's Bathing Beach, Isle of Wight

(Adapted from Donovan, 2019.) Small round holes produced by boring organisms are a common feature of dead shells both on the seashore and in the fossil record (Pickerill & Donovan, 1998). The ichnologist classifies these structures within the ichnogenus *Oichnus* Bromley. Most commonly, but not invariably (Bromley 2004, pp. 466–467), these are the spoor of predatory, boring molluscs, most commonly snails. Big round holes are a feature of fragments of sedimentary beds washed ashore by storms and can be determined as part of a large clavate (club-shaped) boring. These are commonly the product of certain groups of boring bivalve molluscs and assigned to the ichnogenus *Gastrochaenolites* Leymerie. But there are round holes, perhaps a little bigger than typical *Oichnus*, but smaller than common *Gastrochaenolites*, that are found in many of the oyster valves washed ashore on Queen Victoria's bathing beach at Osborne House, East Cowes, Isle of Wight. Their identity is equivocal. Specimens are deposited in the Naturalis Biodiversity Center, Leiden, the Netherlands (prefix RGM).

***Locality***: (See Donovan, 2021, chapter 49.) Queen Victoria's private bathing beach [NGR 525 953] is open to all visitors to Osborne House, East Cowes. It is a rich site for disarticulated oyster valves, other shells and pebbles of Paleogene limestones. Lloyd & Pevsner (2006, figure on p. 211) provided an estate map; the beach is in the area of locality 7 therein ('Landing House') (see also Donovan, 2021, fig. 49.1). The ichnofauna (Donovan, 2014) includes common *Caulostrepsis taeniola* Clarke, *Caulostrepsis* isp., *Entobia* isp. and *Oichnus simplex* Bromley, with rare *Trypanites*? isp. and seagull beak marks.

***Description***: Thirty disarticulated oyster valves of the common European oyster, *Ostrea edulis* Linné, were collected (Figs 4.5, 4.6). Free valves (right, flat) are commoner than attached valves (left valve, convex). Borings are circular, either cylindrical or gently conical, vary in diameter from 1.5 to 2.5 mm and penetrate even the thickest areas of oyster valves. Twenty-three valves are perforated by one small round hole only; others bear two or more holes. Most holes are penetrative; only six valves preserve incomplete borings, including one specimen (RGM.1332432) that also bears a penetrative (complete) boring (Fig. 4.6F). Incomplete borings all start on the outside of the valves. One valve (RGM.1332396; Fig. 4.5F) includes a penetrative boring (in association with *Entobia* isp.) in which a small bivalve is

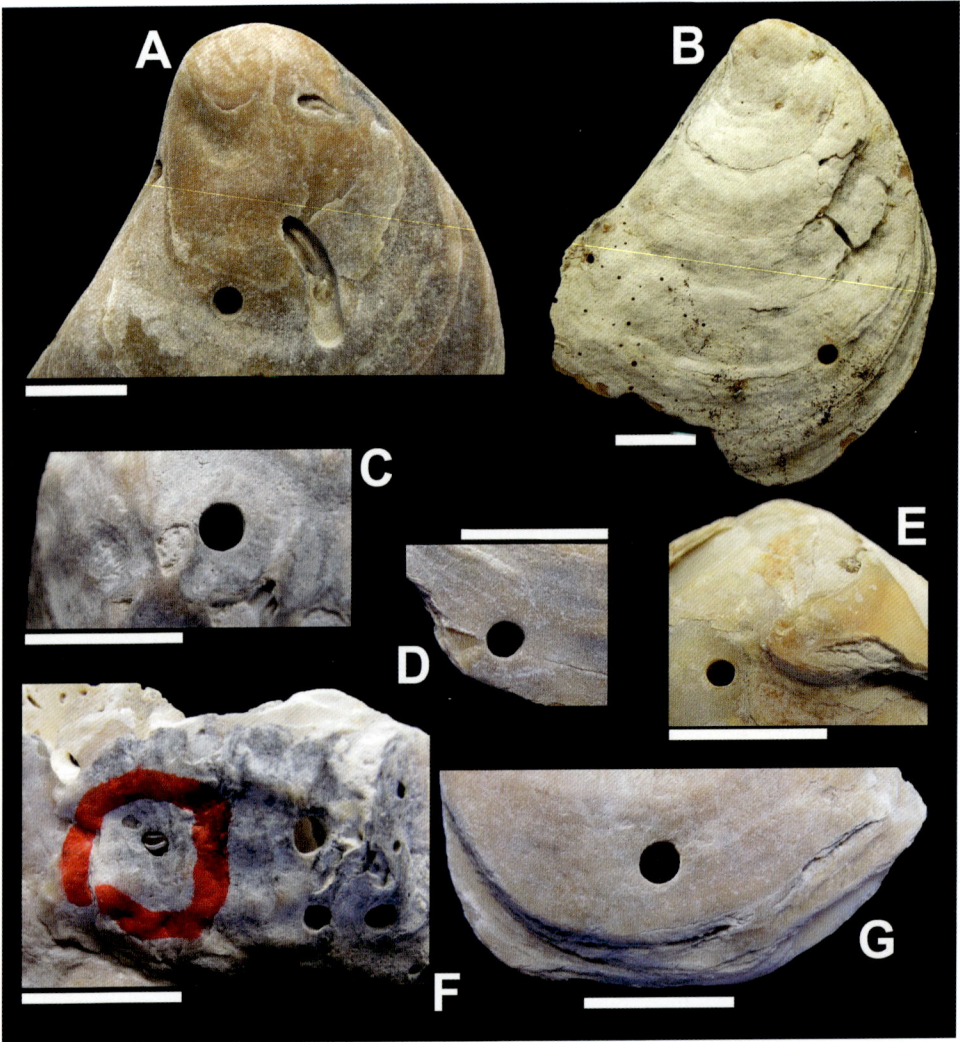

**Figure 4.5** (After Donovan, 2014, fig. 1). *Oichnus simplex* Bromley in *Ostrea edulis* Linné from Queen Victoria's bathing beach, Osborne House, East Cowes, Isle of Wight. (**A**) RGM.1332386, slightly conical boring through moderately thick free valve; *Caulostrepsis taeniola* Clarke to right. (**B**) RGM.1332400, cylindrical boring through thin valve; *Entobia* isp. to lower left. (**C**) RGM.1332389, cylindrical boring through moderately thick part and close to centre of valve. (**D**) RGM.1332387, cylindrical boring close to commissure. (**E**) RGM.1332388[3], slightly conical boring through thick part of valve near umbo. (**F**) RGM.1332396, a cylindrical *O. simplex* (inside red square) with an *in situ* bivalve borer(?) or nestler(?) and in a valve otherwise riddled by *Entobia* isp. (**G**) RGM.1332392, an incomplete boring in a thick part of a valve, near the adductor muscle scar and close to the commissure. Specimens uncoated. All scale bars represent 10 mm.

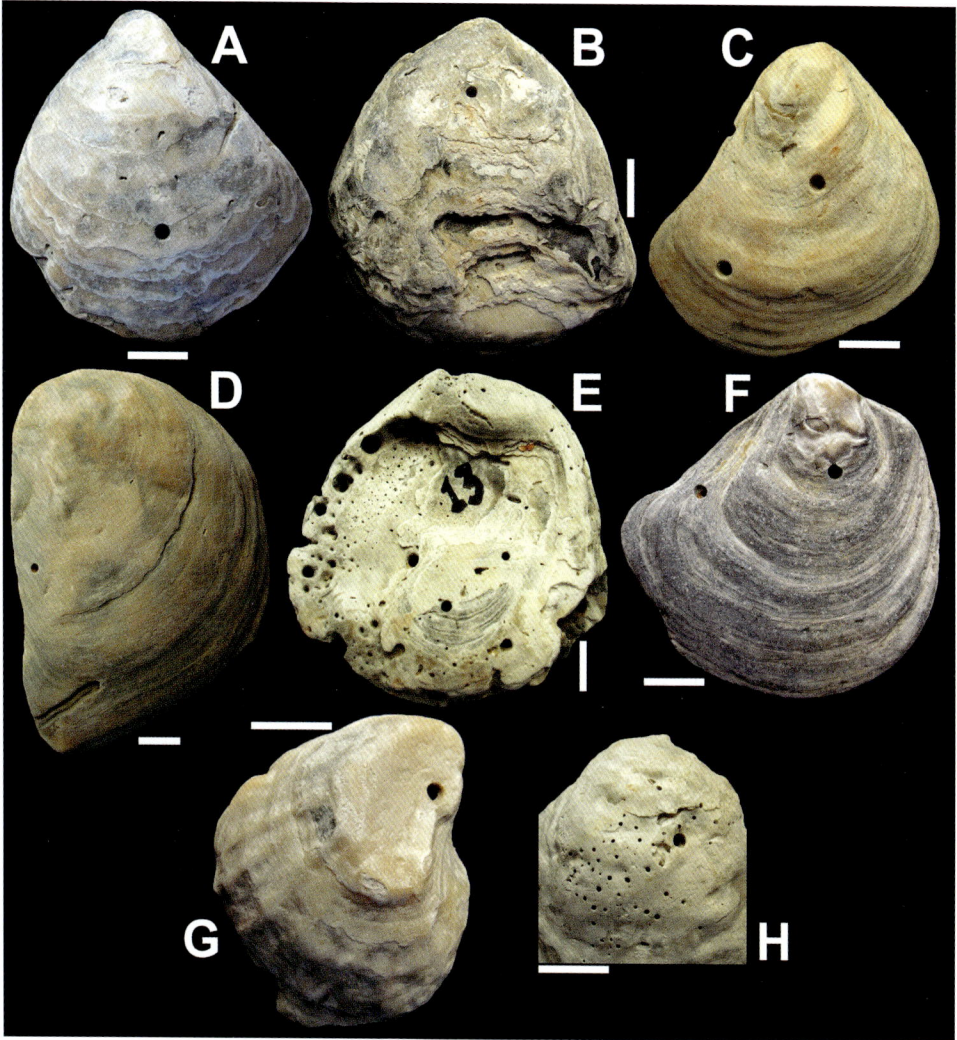

**Figure 4.6** (After Donovan, 2014, fig. 2.) *Oichnus simplex* Bromley in *Ostrea edulis* Linné from Queen Victoria's bathing beach, Osborne House, East Cowes, Isle of Wight. (**A**) RGM.1332391, cylindrical boring through the thickest part of the shell, close to the adductor muscle scar. (**B**) RGM.1332395, incomplete boring near umbo and in thick part of the valve. (**C**) RGM.1332427, two incomplete borings; that in the upper right is almost complete. (**D**) RGM.1332394, incomplete boring; *Caulostrepsis taeniola* towards bottom left. (**E**) RGM.1332398, most, perhaps all of the large boreholes in this valve are *Entobia* isp., but some may be *O. simplex*. (**F**) RGM.1332432, complete (right) and incomplete borings in the same valve; both are in areas where the valve is thick. (**G**) RGM.1332437, slightly conical boring through the attachment scar of an attached valve. (**H**) RGM.1332402, complete, slightly conical boring in close association with a dense infestation of *Entobia* isp. Specimens uncoated. All scale bars represent 10 mm.

A

B

C

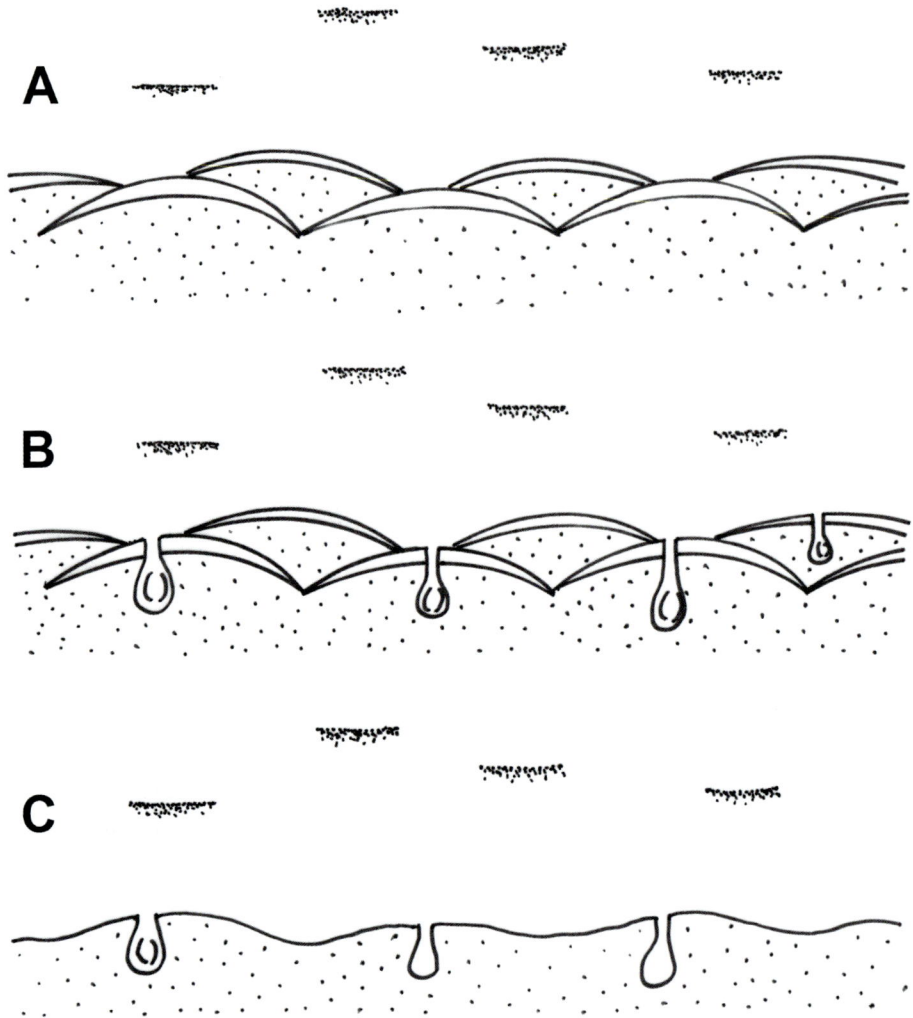

Figure 4.7 (After Donovan, 2014, fig. 3.) Schematic diagram of possible genesis of boring in disartic-ulated valves of *Ostrea edulis* Linné. (**A**) Oyster pavement of numerous dead and disarticulated valves on firm substrate. The valves are dominantly concave-down, which is the most hydrodynamically stable orientation (Brenchley & Newall, 1970). (**B**) A spatfall of larvae of pholadid (boring) bivalves invades the substrate. Pholadids bore into the valves and, where complete, into the substrate (compacted mud?). (**C**) A storm removes the oyster valves and carries them onshore. Some pholadids and their borings are lost; others survive.

preserved. At least one boring (RGM.1332397) is conical and bored from the inside of the valve.

**Discussion**: To reiterate, small round holes in shelly substrates belong to the ichnogenus *Oichnus* Bromley and such cylindrical to slightly conical holes as those discussed herein belongs to the type species, *Oichnus simplex* Bromley. Such is straightforward, but what were the borers and what was their purpose? That is, were they predatory (*Oichnus*) or did each form part of a larger and more extensive domicile (*Gastrochaenolites*)?

Conical borings can be shown commonly to have originated from the external surface, but not invariably; RGM.1332397 was bored from the inside surface outwards. A boring in RGM.1332437 is similarly enigmatic, the broad attachment surface of the valve has been bored (Fig. 4.6G). Six of the valves bear multiple, conspecific borings, but these are mainly incomplete; that is, they provide no evidence for or against predation. Only RGM.1332432 has two borings, one of which is complete (Fig. 4.6F).

The distribution of borings in these oysters are not consistently in any one area of the valve, unlike some assessments of gastropod-generated borings (see, for example, Pickerill & Donovan 1998, pl. 3). Borings are not concentrated in one part of the valve, as might be expected from an efficient predator. Rather, borings are found in both thick and thin parts of the valves. The common, but not invariable association of one valve, one complete boring may be a red herring. That is, it is anticipated that a predator only needs to drill a single boring through which it would feed; however, must solitary boreholes always be predatory? Probably not. Although this would be a typical association of predatory behaviour, and in the absence of evidence for site selection, perhaps they are not due to predatory snails. Rather, they may be domiciles (dominichnia) of small, post-larval boring bivalves. RGM.1332396 (Fig. 4.5F) provides tantalising evidence of a bivalve association with one of these pits, but whether the bivalve is a borer or a post-boring invader, that is a nestler, is impossible to determine without breaking the specimen.

I speculate that these borings included in *O. simplex* actually represent post-mortem, non-predatory borings. If these disarticulated valves lay on the seafloor as a dense accumulation, an oyster pavement (Fig. 4.7A), on a firm, muddy substrate, it is likely that they would have been bored by bivalves. That is, boring bivalves would have treated the valves of dead oyster as part of a lithified seafloor (Fig. 4.7B). Early borings would likely have been narrow, becoming large with time, as the boring bivalves grew in size, but the valves on Queen Victoria's bathing beach were derived from offshore. They may have been transported to their present position in a storm; the boring bivalves may have already been established in the underlying substrate (Fig. 4.7C). Thus, although these holes are undoubtedly assignable to *O. simplex*, if more complete, including the part of the boring in the substrate, it is speculated that they could more correctly be referred to as part of the club-shaped boring, *Gastrochaenolites*. Because most of these oyster valves are thin, they are unlikely to have been infested by bivalves boring into, rather than through them.

# References

Brenchley, P.J. & Newall, G. (1970) Flume experiments on the orientation and transport of models and shell valves. *Palaeogeography, Palaeoclimatology, Palaeoecology,* 7: 185–220.

Bromley, R.G. (2004) A stratigraphy of marine bioerosion. In McIlroy, D. (ed.) *The Application of Ichnology to Palaeoenvironmental and Stratigraphic Analysis.* Geological Society, London, Special Publications, **228**: 455–479.

Donovan, S. K. (2014) Bored oysters and other organism-substrate interactions on two beaches on the Isle of Wight. *Wight Studies: Proceedings of the Isle of Wight Natural History & Archaeological Society,* **28**: 59–74.

———. (2017) Neoichnology of Chalk cobbles from north Norfolk, England: implications for taphonomy and palaeoecology. *Proceedings of the Geologists' Association,* **128**: 558–563.

———. (2019) Round holes in oysters, Queen Victoria's bathing beach, Osborne House, Isle of Wight. *Wight Studies: Proceedings of the Isle of Wight Natural History & Archaeological Society,* **33**: 92–96.

———. (2020) Recent borings in glacial erratics (Carboniferous Limestone), Cleveleys, Lancashire. *The North West Geologist,* **21** (for 2019): 31–49.

———. (2021) *Hands-On Palaeontology: A Practical Manual.* Dunedin Academic Press, Edinburgh.

Donovan, S.K., with Donovan, P.H. & Donovan, M. (2019) (for 2018) A recurrent trinity of Recent borings in clasts around the southern and western North Sea. *Bulletin of the Geological Society of Norfolk,* **68**: 51–63.

Ellis, C. (1968) (first published 1954) *The Pebbles on the Beach.* Faber & Faber, London.

Lloyd, D.W. & Pevsner, N. (2006) *The Buildings of England. The Isle of Wight.* Yale University Press, New Haven, CT.

Murphy, P. (2015) *Exploring the Limestone Landscapes of the Cumbrian Ring.* BCRA Cave Studies Series, **20**. British Cave Research Association, Buxton.

Pickerill, R.K. & Donovan, S.K. (1998) Ichnology of the Pliocene Bowden shell bed, southeast Jamaica. In Donovan, S.K. (ed.), *The Pliocene Bowden Shell Bed, Southeast Jamaica: Contributions to Tertiary and Quaternary Geology,* **35**: 161–175.

Santos, A., Mayoral, E. & Bromley, R.G. (2011) Bioerosive structures from Miocene marine mobile-substrate communities in southern Spain, and description of a new sponge boring. *Palaeontology,* **54**: 535–545.

# CHAPTER 5

## *Aktuo Paläontologie*

An actuopalaeontologist is naturally interested in almost every aspect of marine biology. […] However, he has certain additional interests that are normally not shared by his biological colleagues. (Schäfer, 1972, p. 2)

*Aktuo Paläontologie* examines live and dead organisms as fossils in the making. In the context of this book, shells washed up on the beach may be considered as part of the continuum from life to fossilisation. Dead shells are not fossils, but their preservation may be superior or not; either state begs questions. Dead shells may be examined in the round, whereas many fossils may only be apparent in two dimensions. They can make worthy contributions to many aspects of palaeontology, including preservation, palaeoecology and ichnology.

### Information from dead shells

When examining shells on a beach, the first consideration is obvious, that they must be allochthonous. Dead shells on the beach are not *in situ*, but are derived from the sea, where they lived. Marine invertebrates and fishes die, and are transported offshore, laterally, and onshore by wave and current action. One of my favourite beaches for shell collecting is Southport, north of Liverpool, and on the east coast of the Irish Sea (Donovan, 2021a, pp. 228–234). The best time to collect at Southport is December and January, when winter storms are doing their magic, carrying bioclasts onshore. Shells are more plentiful, more varied and better preserved at this time than during the other ten months of the year. The tide at Southport goes out 2–3 km, yet dead shells make their way to the top of the beach and accumulate thanks to the action of waves.

Even though transported, shells on a sandy beach like Southport may be well preserved. Preservation on a rocky shore is commonly poor. North from Southport is Cleveleys on the Fylde coast (Donovan, 2021a, pp. 200–203; see Chapter 4 herein). The beach at Cleveleys is sandy but also littered with cobbles and pebbles derived from the coastal erosion of boulder clay. There are fewer well-preserved shells than at Southport. The principal factor influencing this disparity is surely the many impacts by shells on stony clasts, which they must crash against when being washed onshore (see also Chapter 22).

I have a research interest in the residence time of dead bivalves before the ligament linking the valves breaks. As a student I was instructed, quite logically, that the soft tissues

**Figure 5.1** Fragment of the posterior part of the razor shell *Ensis siliqua* (Linné) showing the internal (left) and external surfaces (right), specimen NHMM 2020 021 (after Donovan, 2021b, fig. 2) and encrusted, both internally and externally, by the barnacle *Balanus crenatus* Brugiére. The specimen is also encrusted on the internal surface by serpulid worm (probably *Spirobranchus lamarcki* (de Quatrefages)) and bryozoans (*Membranipora*? sp.), but the bivalve's ligament remains intact. Scale in mm and cm.

are lost soon after the death of the mollusc, then the ligament, and thereafter the inner surfaces of the valves (now disarticulated) are available for encrustation by, for example, barnacles and serpulids. My own research has shown that at least some bivalves have a ligament that is tough enough to persist while the inner surfaces of the mollusc's valves are encrusted by shelly invertebrates. While such patterns of infestation are uncommon, rare examples of dead razor shells (*Ensis* spp.) are found with both valves still articulated, and balanids encrusting them both inside and outside (Donovan, 2007, 2021b; Fig. 5.1 herein). But, even on a beach with many hundreds of razor shells, the chances of finding such associations remain small; stay alert.

## Palaeoecology and preservation

A shelly beach is a wonderland for evidence of shelly organisms which interacted in life. There are two principal areas of palaeoecology, autecology and synecology.

**Autecology** is 'the ecology of the individual organism or group, as distinct from that of the whole community'. (Ager, 1963, p. 313)

**Synecology** is 'the ecology of communities, as distinct from that of individual species of groups'. (Ager, 1963, p. 316)

The synecology of allochthonous shell accumulations is obviously problematic. Just because two dead shells are washed up on a beach 50 mm apart is not evidence that they were so closely positioned in life. But shells may be preserved in close association with certain organisms that they lived with in life, most obviously by those that encrusted them or bored into them. One example that is simple to recognise and interpret is a contemporary borehole made in a molluscan substrate, ancient or modern (see Fig. 5.2 for an ancient example). Consider, for example, a mussel valve encrusted externally by balanids and pierced by a single, conical boring. This trace, *Oichnus paraboloides* Bromley, is typically the spoor of predatory naticid snails. The boring tells us how the mussel died. Further, the mussel was an epifaunal bivalve, so it was probably encrusted by balanids in life; there no encrusters on the inner surface of the valve.

The autecology of an organism or group of specimens of one species found washed up on the beach is best considered as a study in functional morphology. That is, consider how the preserved hard parts of the organism worked together as an indicator of how it lived.

**Figure 5.2** Belemnites in beach clasts (after Donovan, 2022, fig. 3), *Belemnitella*? sp., north Norfolk coast between Overstrand and Cromer (see Chapter 20). Chalk, Upper Cretaceous. (**A**, **left**) NHMM 2021 030, partial belemnite rostrum preserved in calcite. (**B**, **right**) NHMM 2021 029, external mould of partial belemnite rostrum preserving a dense infestation of borings, *Trypanites*? sp., in flint. Specimens uncoated. Scale in mm and cm.

Consider, for example, the differences in form between infaunal burrowing and epifaunal bivalve molluscs. Burrowing bivalves are equivalve, each valve the same size and a mirror image of the other. This gives the shell a streamlined shape, favouring its movement through the sediment. Internally, the pallial sinus (see Chapter 13) demonstrates that the siphons could be withdrawn into the shell. The deeper the pallial sinus, the larger the siphons and the deeper the bivalve was able to burrow.

In contrast, epifaunal bivalves are commonly inequivalve, have no pallial sinus and may have been attached in life. Mussels are near-equivalve (but not quite) and attach by byssal threads (Chapter 13). Scallops have a flattened valve and may attach by a byssus or can be free-swimming. Oysters are commonly very irregular, with a deep attached (cemented) valve and an irregularly flattened free valve.

## Traces old and new

(Adapted from Donovan & Fearnhead, 2014.) Modern traces – what they are and what they are not – are of broad interest and relevance. Traces are all around us, but they are such a part of the environment that they rarely engender comment. Go for a walk on a sandy beach or in the snow and you will leave your tracks, as do other animals. Pick up a handful

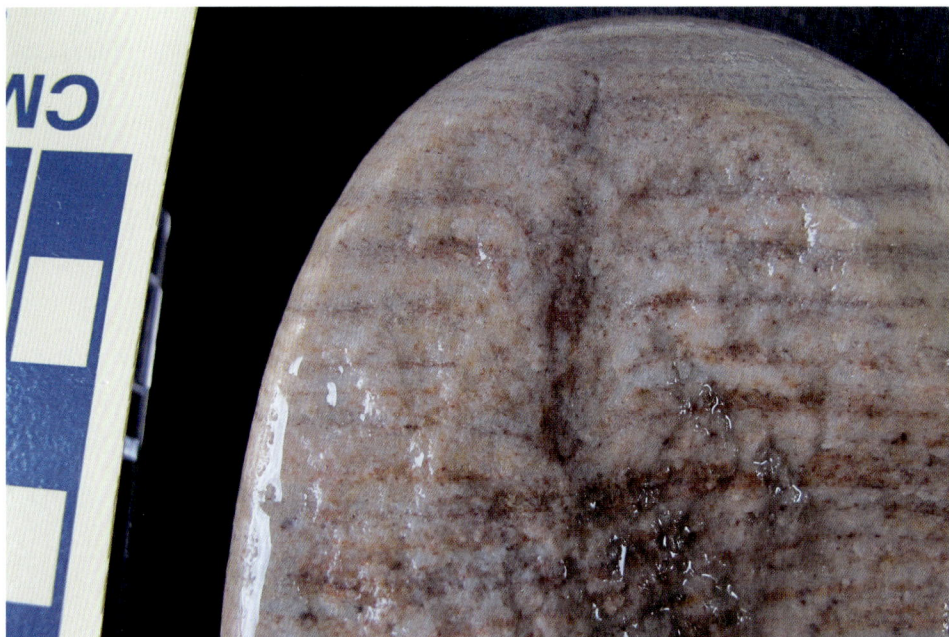

**Figure 5.3** (After Donovan, 2011, fig. 3B.) A burrow found on a beach, but, and unexpectedly, not a modern trace, but a trace fossil preserved in an erratic sandstone clast (Pennsylvanian?). The vertical burrow *Monocraterion* isp., disturbing horizontal beds as the producer dug upwards to escape from rapid burial (specimen Naturalis Biodiversity Center, Leiden, the Netherlands (prefix RGM 617 925)), from the coast between Overstrand to Cromer, north Norfolk (Chapter 20). Scale in cm.

**Figure 5.4** (After Donovan, 2011, fig. 2C.) Part of a Recent wide, but incomplete, club-shaped boring in chalk, *Gastrochaenolites* isp. (RGM 617 924), most likely produced by a boring bivalve. From the coast between Overstrand to Cromer, north Norfolk (Chapter 20). Scale in cm.

of shells from the strandline and chances are that one or more will bear a small, conical, circular boring, evidence of predation by a naticid gastropod (see above).

A trace is 'frozen' biological activity, the result of an organism doing something and then moving on. The organism isn't there any more, unless you fortuitously catch them in the act, but they have left evidence of their activities: 'Kilroy was here'. We, and an array of organisms, make macro-traces that we can note if we look for them. A trace fossil is the product of an ancient organism and is evidence of prehistoric activity by ancient plants and animals. The study of trace fossils is ichnology; that of modern traces, neoichnology.

Traces can take many guises. Those most commonly encountered are trails, tracks and trackways, burrows and borings. Trails are produced by an organism dragging itself over a sedimentary surface and leaving a groove behind. Trails are more commonly produced by organisms on or just under the seafloor that were capable of moving about, but which lack legs, such as a gastropod or annelid worm moving over a muddy bed. On the beach, these may be produced at high tide; they will be destroyed (and replaced) at the next high tide. They are not collectable but take a camera to preserve a photographic 'trace' of a trace.

I am evangelical about traces and trace fossils; they are important sources of data for ecology and palaeoecology but are too easily ignored. A track is a footprint; a succession of tracks is a trackway. An organism needs discrete walking limbs to make a track or trackway, which limits the range of potential producers. That is, they are commonly the product of only two groups – arthropods and tetrapods. On a beach, the most likely trackway producers amongst the arthropods are the crabs.

Burrows take us into the sediment, the land of the digger. Burrowers will be most active on a beach at high tide; to examine them at low tide will need a spade (but see an ancient example, collected from a beach, in Fig. 5.3).

Borers are cutters, grinders and dissolvers of hard substrates, such as shells, wood and limestones. Some are predators or parasites, whereas others inhabit a protected environment (Fig. 5.4).

## References

Ager, D.V. (1963) *Principles of Paleoecology.* McGraw-Hill, New York.

Donovan, S.K. (2007) A cautionary tale: Razor shells, acorn barnacles and palaeoecology. *Palaeontology,* **50**: 1479–1484.

———. (2011) Aspects of ichnology of Chalk and sandstone clasts from the beach at Overstrand, north Norfolk. *Bulletin of the Geological Society of Norfolk,* **60** (for 2010): 37–45.

———. (2021a) *Hands-On Palaeontology: A Practical Manual.* Dunedin Academic Press, Edinburgh.

———. (2021b) Fast post-mortem encrustation of razor shells: Examples from the Irish Sea and palaeontological implications. *Proceedings of the Yorkshire Geological Society,* **63**: 301–304.

———. (2021c) Taphonomy of fossil invertebrates in flint beach clasts (Upper Cretaceous), north Norfolk coast. *Bulletin of the Geological Society of Norfolk,* **72**: 3–10.

Donovan, S.K. & Fearnhead, F.E. (2014) The nature of trace fossils. *Deposits,* **39**: 38–43.

Schäfer, W. (G.Y. Craig, ed.). (1972) *Ecology and Palaeoecology of Marine Environments.* University of Chicago Press/Oliver & Boyd, Chicago and Edinburgh.

# CHAPTER 6

# Health and safety

The concept of health and safety has a bad name (Donovan, 2021, pp. 58–60). For professional field scientists, it means more paperwork to summarise much of what must be obvious. For most readers of this book, amateurs and students alike, there is no paperwork, but the responsibilities are much the same. Do everything in your power to ensure a safe day on the beach and act in a sensible manner once you are there. Accidents and slips of various sorts may occur, but they can be kept to a minimum by thought beforehand and care on the day. Health and safety must be a state of mind for the field scientist, an awareness of your environment. It was ever so before the phrase health and safety was first coined by a civil servant in a nameless committee.

## The beach

I presume that your excursions are either just you or a small group of friends and these comments are written from this point of view. If you are taking a large, organised group, consult the Foreshore Code of Conduct (see below). Beaches are commonly friendly environments, but they can also be dangerous. Man-made items like barbed wire, discarded syringes and unexploded ordnance may all occur locally. The natural environment may also be less than welcoming.

- Dry, compact clay is solid, but potentially slippery, and will support you; wet, muddy beaches are not favourable for the walker, slowing you down by sucking at your boots. Dawdle too long at any given point and you may be trapped, your boots sinking too deep into the beach for easy removal. If you are going to stand in one spot for any length of time, stand on a wooden board or plank to spread your weight.

- Sandy beaches provide good surfaces for the walker except where the sand is dry, slowing you down with each step, or 'quick' through being too wet. Unlike the heroes in adventure movies who seem unable to think their way out of quicksand (I am looking at you, Indiana Jones), reversing direction to stand on firm ground once again is the obvious solution.

- Rocky beaches are different. They involve climbing and scrambling, often on surfaces polished by wave action and made slippery by algae. There is thus a falling hazard, with the chances of broken bones and severe bruising; it was such a fall in the field that led to the death of one of my geological heroes, C.T. Trechmann (Donovan, 2003). A walking stick may help, giving you three legs rather than two, but the chance of a sliding fall is still there. Going on all fours is helpful in some places. Best of all, stay off wet, slippery rocks, although you can still fall when they are dry.

Do not forget the emergency supplies, medical and otherwise. You ought to make sure that they are adequate but should be kept manageable. They are necessary and can be invaluable, but weight and bulk should be kept to a minimum. I am diabetic and I make sure to take all necessary medication for the day with me. A bottle of water is always useful for drinking or washing wounds (or specimens). A small first aid kit, such as some sticking plasters and a tube of antiseptic cream will be needed sooner or later. Some long-lasting snacks are desirable, such as cereal bars, chocolate and/or a bag of raisins. I prefer raisins to fresh fruit, which are squashy and soon go mouldy if forgotten; for example, I have a reputation amongst some co-workers as a man who sits on bananas. In contrast, you can leave all of the 'dry goods' in your backpack for months and they will still be edible when eventually needed. A diabetic such as your author keeps a tube of Dextrose tablets in his backpack, as well as my medication. I also keep painkillers, such as Paracetamol, and Rennies to hand. Sun cream and insect repellent are light to carry and may be essential. I normally wear a hat with a broad brim. The hat keeps the sun out of my eyes, the rain off my glasses and my head warm. In cold seasons, extra gloves, socks and scarves are light and potentially useful.

Do you know where you are? Do not get lost. Have a good topographic map of your field area (1:50,000 scale at least), a GPS which you know how to use, and a field guide, if available. Maps have the advantage over a GPS in that there are no batteries to go flat, but you should also have spare batteries. Similarly, make sure your mobile phone has an adequate charge, although a lack of signal on a cliffed coast is not unexpected.

Various codes of conduct for the field geologist are available online. I recommend the Geologists' Association Code for Fieldwork. This is available at the Association's website under Publications, along with a Foreshore Code of Conduct; it covers beach behaviour in greater detail than I do herein.

## Onshore and inland

The sea rises and falls with the tides; hopefully the inland side of the beach is more stable and accessible but beware of rockfalls and landslides. Most importantly, know where you are. There is little need to worry with a low sea wall in a holiday resort with numerous access points. Away from such safe environs there are obstacles such as cliffs to stop your withdrawal from a rising tide. If you need to beat a hasty retreat, where is your point of egress? Preferably you will have at least two potential exits close at hand, the way you came onto the beach and another in the direction that you are travelling. Unless scaling unstable cliffs is your expertise, do not contemplate it. Before getting onto the beach, consult the tide tables and know when to vacate the strand. Do not expect an easy alternative out to sea; stay on *terra firma* and keep your feet dry.

## References

Donovan, S.K. (2003) Charles Taylor Trechmann and the development of Caribbean geology between the wars. *Proceedings of the Geologists' Association*, **114**: 345–354.

———. (2021) *Hands-On Palaeontology: A Practical Manual.* Dunedin Academic Press, Edinburgh.

# II

# What to look for

# CHAPTER 7

## Provenance

**provenance** [...] **1.** The place of origin, derivation, or earliest known history, esp. of a work of art, manuscript, etc. (*New Shorter Oxford English Dictionary*, 1993, p. 2392)

This second part, 'What to look for', is largely self-explanatory. The chapters are guided by my own experiences on the beach. The focus is shelly invertebrates and, to a lesser extent, their traces. I have avoided unmineralised invertebrates like jellyfishes, which do have a fossil record, although their ancestors are neither common nor diverse in the rock record. I also avoid the vertebrates, which are not uncommon on beaches, ranging from stranded fishes to beached whales. They already have an ample literature, and I prefer to leave these back-boned show-offs to the experts. I prefer to tread the record left by the spineless wonders.

Herein, I preface Part II by examining the question that every collector needs to ask of every erratic specimen, where does it come from? Determining provenance is a study that underlies all of our beach studies. In some instances, this is easy to determine, such as the invertebrate trails on a sandy or muddy beach. Very obviously, such modern traces are ephemeral, preserved *in situ* during or after the last falling tide and waiting to be reworked by the next incoming tide. Preserving them is the province of the camera.

I divide this chapter into two sections, examining the provenance of modern shells and ancient fossils on the beach. The latter obviously poses two separate questions: how did the fossil get into the rock; and how did the rock end up on the beach? Only the second question should really be confronted in the present study, but the two may be interlinked, as I shall demonstrate.

### Provenance of shells on the beach

The obvious answer is from the sea, but this is only a partial, albeit correct response as far as it goes. The majority of shells on the beach are carried there by traction or saltation, rolling onshore thanks to bottom currents. The best time to collect ample shells on the beach is after a major storm, when erosion of the nearby shallow seafloor will liberate many dead (and living) infaunal invertebrates for transport onshore. For example, after spring storms, the beach at Zandvoort aan Zee (Donovan, 2025) is littered with many thousands of shells, dominated by the razor shell *Ensis* (Donovan, 2011a). Any live shell so disinterred and swept on the beach will quickly be a feast for hungry shorebirds. Indeed, whatever time of year you collect, it is the major storms that will bring most benthic shells onshore.

Other shells may float ashore, particularly those of cephalopods (Chapter 14). The internal shell of the vagile cuttlefish *Sepia* is a common sight on beaches around the British Isles. Dead cuttlefish release their shells after the soft tissues have rotted away. The shell may float for weeks or months until either washed onshore (Jongbloed *et al.*, 2016; Figs 14.2, 19.8 herein) or eventually sink after becoming waterlogged. There is also the chance that the carcass of a benthic organism may fill with decomposition gasses and float ashore. This has been reported for echinoids (Reyment, 1986); such occurrences are referred to as nekroplankton. Other benthic shelly organisms such as barnacles or serpulids may have attached to floating objects (wood, shells of dead cephalopods, plastics, bottles), forming the pseudoplankton (Wignall & Simms, 1990; Fig. 7.1 herein) and, again, may eventually be carried onshore.

Some shelly invertebrates live on the beach. Rock pools can support a range of shelly and soft-bodied invertebrates that may change after each high tide. There are also burrowing bivalves that live in a beach environment. I particularly mention the Antillean bivalve *Donax denticulata* Linné, which is a common intertidal burrower on Farquhar's Beach in south-central Jamaica (Donovan & Miller, 1999, p. 36). As the tide comes in, *Donax* migrates up the beach; as the tide retreats, *Donax* follows it down. If you watch the waves

**Figure 7.1** (After Donovan, 1999, fig. 1; 2011b, fig. 2.) Exotic pseudoplankton. A cobble of pumice encrusted by *Lepas anatifera* Linné that floated onto the Palisadoes, south coast of Jamaica, West Indies. Larger, more mature individuals occur towards the circumference, while a few minute juveniles are discernible more centrally. Field of view about 82 mm wide.

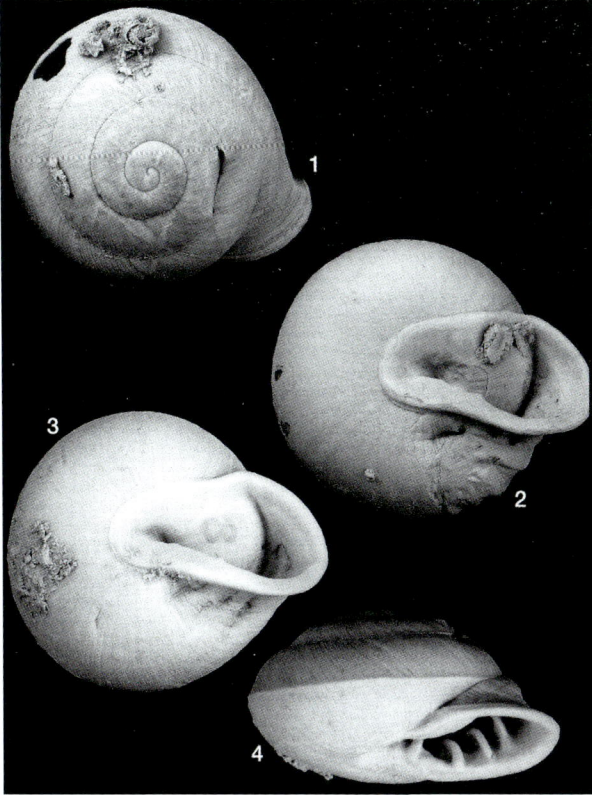

**Figure 7.2** (After Paul & Donovan, 2006, pl. 12.) Large land snails of the sort that may be washed onto a beach or into deep water. *Pleurodonte sinuata* (Müller), Late Pleistocene, Red Hills Road Cave, parish of St Andrew, Jamaica. Specimens are deposited in the Naturalis Biodiversity Center, Leiden, the Netherlands (prefix RGM). (**1**, **3**, **4**) RGM 188 752. (**1**) Apical view, about 21.5 mm wide. (**3**) Umbilical view. (**4**) Apertural view. (**2**) RGM 188 753, umbilical view, about 25.7 mm wide.

swash up and down the beach, *Donax* will be seen emerging from the sand and being transported by the moving water.

Confusion may be caused by the shells of dead land snails on the beach. That land snails are preserved in deep water deposits is well known, such as the Pliocene Bowden Formation of Jamaica (Goodfriend, 1993) and the Miocene Grand Bay Formation of Carriacou (Jung, 1971). How did they get there? I have seen terrestrial *Pleurodonte* snail shells (Fig. 7.2) rolling in a stream over a Jamaican beach towards the sea following a rainstorm. *Pleurodonte* is a large, robust shell that occurs in both the named deposits; indeed, its only pre-Pleistocene fossil record is in marine deposits. So, beware that a beach may preserve an allochthonous mixture of shells from both offshore and adjacent terrestrial environments.

## Provenance of fossils on the beach

As with shells, so with fossils. Many of the comments regarding shells on the beach are also applicable to rocks and fossils. They may be derived from inland, the beach itself or offshore. One key factor is longshore drift, whereby shells and rocks may be transported along a coast by asymmetrical wave action. Clasts are carried up the beach and dragged down again, but along different tracks driven by oblique wave actions (Pilkey *et al.*, 2011, pp. 88–89). For example, one of my favourite beaches for collecting rock clasts is at Cleveleys in Lancashire,

**Figure 7.3** (After Donovan, 2020, fig. 1.) Provenance of cobbles. A view north from high on the beach at Cleveleys, Fylde, Lancashire. The outline of the peaks of the Lake District is plainly visible; it is a sunny, near-windless day in late December and the Irish Sea is particularly calm. The cobbles on the beach are obvious and diverse, dominated by lithologies derived from the Lake District. This is a marvellous place for geological beachcombing.

north of Blackpool (Donovan, 2021, pp. 200–204; Fig. 7.3 herein). It is a fine site to find rock clasts of many lithologies, most presumed to have been derived from the Lake District, the peaks of which can be seen in the distance, and reworked out of Pleistocene 'boulder clay'. But there is no exposure of 'boulder clay' on the beach. The hundreds of clasts that now populate the upper part of the beach had to come from somewhere else, probably both from offshore and longshore transport from eroding 'boulder clay' further to the north.

Recognising that clasts may be transported laterally, mass movement of rocks and unlithified sediment downslope – the many forms of landslides – may replenish a beach in both the immediate area and laterally by longshore drift. Landslides are known to supply fossils to many beaches around Britain, such as the Cretaceous Gault Clay Formation at Folkestone, Kent (Hadland, 2018) and the Paleogene London Clay Formation on the Isle of Sheppey, Kent (Pitcher, 1967). But these rock units are soft clays from which lithified fossils may easily be winnowed. What of the Late Pleistocene Farquhar's Beach red beds on the south coast of Jamaica (Donovan *et al.*, 2010) from which huge boulders litter the beach? This sedimentary deposit lies on an angular unconformity over Neogene siliciclastics and limestones of the August Town Formation. The red beds, with land snails and tree roots, enclose limestone boulders derived from the Miocene Newport Formation, a marine fossiliferous limestone succession, exposed nearby and inland. Thus, as the huge boulders of the Farquhar's Beach red beds break down mechanically, they release clasts of both fossiliferous marine limestones and fossiliferous terrestrial red beds into the beach sediment budget.

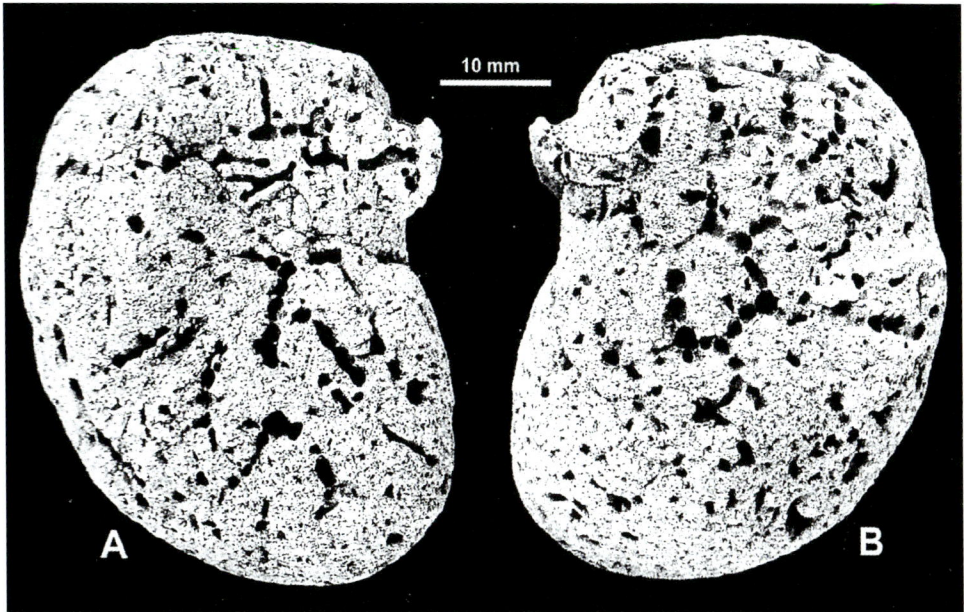

**Figure 7.4** (After Donovan & Lewis, 2010, fig. 1.) Sponge borings ancient and modern meet. Two surfaces of a bored chalk pebble, RGM 544 413, from the beach at Overstrand, north Norfolk. (**A**) Slightly concave surface, belemnite top right and in transverse section. (**B**) Convex surface, belemnite top left. Outline of the belemnite is picked out by pecked lines. Borings in the belemnite are both modern (open) and ancient (infilled by lithified chalk).

Cobbles on the beach may preserve evidence of complex interactions between fossil and Recent organisms, separated by millions of years. Fossil shells may be modified by modern borings (Donovan *et al.*, 2014) or provide substrates for extant encrusters. My favourite example is a reworked Upper Cretaceous belemnite, bored by sponges (*Entobia* ispp.) immediately after the belemnite's demise and again in the recent in a near-beach environment (Donovan & Lewis, 2010; Fig. 7.4 herein).

# References

Donovan, S.K. (1999) Pumice and pseudoplankton: Geological and paleontological implications of an example from the Caribbean. *Caribbean Journal of Science*, **35**: 323–324.

———. (2011a) Post-mortem encrustation of the alien bivalve *Ensis americanus* (Binney) by the barnacle *Balanus crenatus* Brugière in the North Sea. *Palaios*, **26**: 665–668.

———. (2011b) Beachcombing and palaeoecology. *Geology Today*, **27**: 25–33.

———. (2020) Recent borings in glacial erratics (Carboniferous Limestone), Cleveleys, Lancashire. *The North West Geologist*, **21** (for 2019): 31–49.

———. (2021) *Hands-On Palaeontology: A Practical Manual*. Dunedin Academic Press, Edinburgh.

———. (2025) A beachcomber's field guide to the other side of Doggerland: Zandvoort aan Zee, Noord Holland, The Netherlands. *Bulletin of the Geological Society of Norfolk*, **75**: 45–55.

Donovan, S.K., Blissett, D.J. & Jackson, T.A. (2010) Reworked fossils, ichnology and palaeoecology: An example from the Neogene of Jamaica. *Lethaia*, **43**: 441–444.

Donovan, S.K., Harper, D.A.T., Portell, R.W. & Renema, W. (2014) Neoichnology and implications for stratigraphy of reworked Upper Oligocene oysters, Antigua, West Indies. *Proceedings of the Geologists' Association*, **125**: 99–106.

Donovan, S.K. & Lewis, D.N. (2010) Notes on a Chalk pebble from Overstrand: ancient and modern sponge borings meet on a Norfolk beach. *Bulletin of the Geological Society of Norfolk*, **59** (for 2009): 3–9.

Donovan, S.K. & Miller, D.J. (1999) Report of a field meeting to south-central Jamaica, 23 May 1998. *Journal of the Geological Society of Jamaica*, **33** (for 1998): 31–41.

Goodfriend, G.A. (1993) The fossil record of terrestrial molluscs in Jamaica. In Wright, R.M. & Robinson, E. (eds) *Biostratigraphy of Jamaica. Geological Society of America Memoir*, **182**: 353–361.

Hadland, P. (2018) *Fossils of Folkestone, Kent*. Siri Scientific Press, Manchester.

Jongbloed, C.A., Gier, W. de, Ruiten, D.M. van & Donovan, S.K. (2016) Aktuo-paläontologie of the common cuttlefish, *Sepia officinalis*, an endocochleate cephalopod (Mollusca) in the North Sea. *PalZ*, **90**: 307–313.

Jung, P. (1971) Fossil mollusks from Carriacou, West Indies. *Bulletins of American Paleontology*, **61** (269): 147–262.

*The New Shorter Oxford English Dictionary*. (Brown, L., ed.) (1993) 2 vols. Clarendon Press, Oxford.

Paul, C.R.C. & Donovan, S.K. (2006) Quaternary land snails (Mollusca: Gastropoda) from the Red Hills Road Cave, Jamaica. *Bulletin of the Mizunami Fossil Museum*, **32** (for 2005): 109–144.

Pilkey, O.H., Neal, W.J., Kelley, J.T. & Cooper, J.A.G. (2011) *The World's Beaches: A Global Guide to the Science of the Shoreline*. University of California Press, Berkeley.

Pitcher, W.S. (1967) Itinerary VI: North Kent coast – Isle of Sheppey: Warden Point. In Pitcher, W.S., Peake, N.B., Carreck, J.N., Kirkaldy, J.F. & Hancock, J.M., *The London Region* (*South of the Thames*). rev. ed. *Geologists' Association Guides*, **30B**: 25–26.

Reyment, R.A. (1986) Nekroplanktonic dispersal of echinoid tests. *Palaeogeography, Palaeoclimatology, Palaeoecology*, **52**: 347–349.

Wignall, P.B. & Simms, M.J. (1990) Pseudoplankton. *Palaeontology*, **33**: 359–378.

# CHAPTER 8

## Corals and sponges

### Form and function

In this chapter, I 'lump' together two groups of unrelated colonial organisms that are important in the fossil record and at the present day, but which are not necessarily common in the beaches of northern Europe except as reworked fossils. In and around the British Isles, the cnidarians, which include the corals, may be common as unmineralised, solitary organisms on the beach, such as sea anemones in rock pools and jellyfish washed up by the tide. But in the humid tropics, where there may be coral reefs offshore, the beach can be littered by countless clasts of coral washed onshore by major storms. Evidence of sponges is commonest in bored clasts perforated by the clionaids.

**Cnidarians**: The cnidarians, otherwise called the coelenterates, are the group which includes jellyfishes, sea anemones, corals, hydroids and the Portuguese man-of-war. The basic body plan is simple and is essentially a sac-like stomach with a single opening, surrounded by a cluster of tentacles which are adapted mainly for food capture and defence (Fig. 8.1).

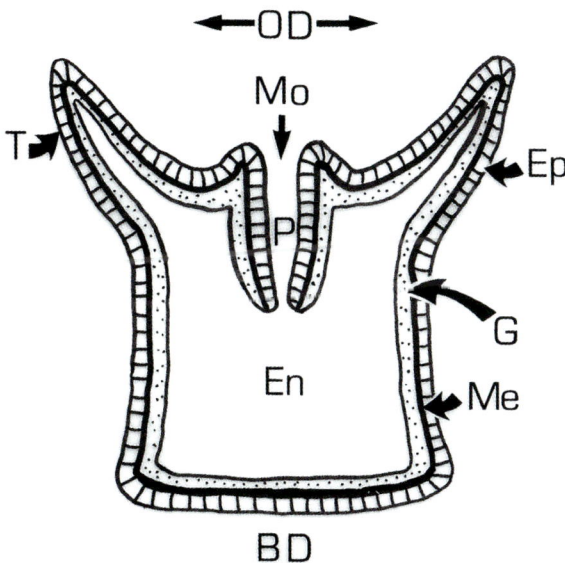

**Figure 8.1** Schematic cross-section through an anthozoan (coral) polyp with a pharynx (redrawn after Oliver & Coates, 1987, fig. 11.2C). Key: BD = basal disc; En = enteron cavity; Ep = epidermis; G = gastrodermis; Me = mesogloea; Mo = mouth; OD = oral disc; P = pharynx; T = tentacle

Cnidarians take two forms with similar body plans, either of a free-living medusoid, such as a jellyfish, or an attached polyp (Fig. 8.1), such as a sea anemone or coral. The body wall comprises two layers of cells, the outer ectoderm or epidermis and the inner endoderm or gastrodermis, separated by a jelly-like partition called the mesogloea. The body wall encloses the gastrovascular or enteron cavity, in which food is digested. The only opening to this cavity is called the mouth, although this must also function as the anus. In polyps, the mouth is directed upwards (Fig. 8.1) and in medusoids it is oriented downwards. The mouth is surrounded by a cluster of tentacles, which capture food and bring it to the mouth. Complex responses, such as the capture of prey using a battery of stinging cells, are co-ordinated by a nerve net. Reproduction is either asexual ('budding') or sexual. Budding occurs by the branching off of new individuals from the parent organism.

Preservation in the fossil record is favoured by hard parts. Jellyfishes (medusoids) have a fossil record that extends back to the late Precambrian and other soft-bodied cnidarians are known from rare fossil occurrences. However, it is the stony corals that have the best fossil record amongst the cnidarians. The corals and sea anemones belong to the Class Anthozoa of the Phylum Cnidaria. The fossil corals are known from three major (and some minor) groups: Order Rugosa and Order Tabulata (both Palaeozoic); and Order Scleractinia (Middle Triassic to Recent).

The scleractinians (or hexacorals) are differentiated from the nominally similar Palaeozoic rugose corals principally on their pattern of insertion of septa, in groups of six rather than four (Fig. 8.2); in having an aragonitic, rather than calcitic, corallum; and on their age range. The skeleton is formed from tubular, cup-like structures, either one (solitary coral) or more than one (colonial), although the form will be fixed in any particular species.

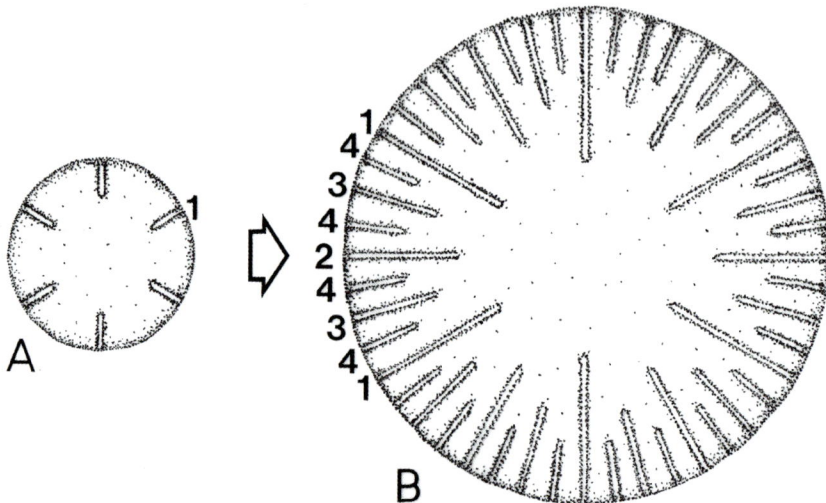

**Figure 8.2** Schematic diagram of the pattern of septal insertion in a scleractinian corallite. (**A**) Simple insertion of six primary septa only, such as found in an early stage of growth. (**B**) More mature growth stage with insertion of 4 orders of septa, that is, 6 primary septa (1), 6 secondary septa (2), 12 tertiary septa (3) and 24 quaternary septa (4).

Individual 'tubes' in a colony are termed corallites; the entire colony is called the corallum. The corallum can form an open structure or may be dense. The wall of the tube is the epitheca and the cup-like structure is the calice. In a living scleractinian, the soft tissues of the polyp sit in the calice and overlap the epitheca. The septa (singular, septum) are radial, platy elements within the corallites which supported folds (mesentaries) of the gut. The pattern of septal insertion is illustrated (Fig. 8.2). Dissepiments are concentric structures which connect the septa. Growth involves secretion of new epithecal wall which lengthens the tube beneath the polyp.

Colonial scleractinians are the reef-building corals (hermatypic) and are distinguished from the solitary, ahermatypic corals. Coral reefs are best known from shallow-water depths in tropical regions, but deep, cold-water coral reefs also exist. Tropical, shallow-water reefs are limited to the top 100 m or so of the sea, known as the photic zone, that is, the region that is penetrated by sunlight. These shallow-water reef builders have a symbiotic relationship with algae that live within the endodermal cells of the coral. Here, the algae produce oxygen and carbon dioxide, which the coral utilises in respiration and secretion of the corallum, respectively, while the algae have a protected environment. Corals with algae are termed zooxanthellate; corals without algae are non-zooxanthellate.

The corallum of colonial scleractinian corals is attached and composed of a number of corallites, each topped by a cup-like calice in which the living polyp is situated. The early colony forms when a soft-bodied larva settles and secretes a thin, calcareous basal plate. As the polyp grows, an epitheca is secreted. A colony is built up by successive budding of the corallum. Colonial corals have taken a number of forms within the limitations of their sessile benthic existence (see, for example, Benton & Harper, 1997, fig. 6.6; Clarkson, 1998, fig. 5.5). The following descriptive terms refer to different forms of scleractinian coral colonies.

**dendroid** – composed of irregular branches.

**phaceloid** – corallites more or less parallel and sometimes joined by connecting processes.

**thamnasteroid** – septa of adjacent corallites confluent (run together to form a single structure) and are often sinuous.

**aphroid** – septa are reduced at their outer ends so that neighbouring corallites are united by a zone of dissepiments.

**cerioid** – walls of adjacent polygonal corallites are closely united and originate from the dissepiments or septa, not from the epitheca.

**plocoid** – corallites have separated walls but are united by dissepiments.

**ramose** – encrusting colonies.

**meandroid** – corallites are arranged in linear series without cross-walls but are confined within the lateral walls which run irregularly over the surface like convolutions of a human brain.

**hydnophoroid** – centres of the corallites are arranged around little 'hillocks' called monticules.

Zooxanthellate corals are major reef builders in the tropics, preferring open marine conditions with normal salinity sea water, high temperatures and shallow water depths (optimal conditions are 15 m or less depth and $25^0$ to $29^0$ C). Modern reefs are best developed in the Indian and Pacific Oceans, and the Caribbean Sea. The Caribbean and Pacific faunas were similar during the Paleogene and Miocene, as there was an open seaway between North and South America, linking the regions and allowing interchange of larvae. Since the late Pliocene, this seaway has been closed by the emergence of the Isthmus of Panama and the Caribbean fauna has developed in isolation.

**Porifera**: The sponges are multicellular, sessile and aquatic, mainly marine, but sometimes freshwater. There is commonly a skeleton of either calcareous or siliceous spicules, or of fibres of spongin, a horny, organic material. Sponge structure is essentially based on a series of water canals. Attachment is at the base. Nutrition is on fine detritus and plankton captured by choanocyte cells. Asexual reproduction occurs by budding and there is unique embryonic development of larvae following sexual reproduction.

The sponges are the most primitive group of multicellular animals and are not metazoans. They have reached the cellular level of organisation, that is, they have cell differentiation without co-ordination into tissues. There is no nervous system or sensory cells.

There are two layers of cells in the body wall, an inner choanoderm (mainly composed of collared, flagellate, choanocyte cells) and an outer pinacoderm comprised of flattened epithelial cells called pinacocytes. The primitive shape of a sponge is a vase-like structure (Fig. 8.3). Water enters a sponge through numerous pores, formed from tubular porocyte cells, and passes out through a single exhalent osculum. The body wall surrounds a water-filled cavity, the spongocoel. This structure imposes size limitations on sponge development. The rate of water flow is slow, because the large spongocoel contains too much fluid for it to be moved out of the osculum rapidly. The inner choanoderm layer moves the water using its flagella. However, an increase in size produces a cubic increase in volume, but only a square increase in choanoderm area. This problem was overcome during the evolution of the sponges by folding the body wall, which increased the surface area of the choanoderm; and reducing the relative size of the spongocoel, which decreased the volume of water to be circulated. The choanocytes do not line the spongocoel in more advanced poriferans but instead are confined to the flagellated canals. There are also distinct incurrent canals. Although normally small, such sponges may reach several hundred mm in height. In some groups, further folding occurred so that the flagellated canals form small, rounded chambers. The spongocoel disappeared apart from water channels leading to the osculum. This is the most efficient morphology of sponge and may reach large sizes.

Growth is generally irregular, with branching (Fig. 8.3), encrusting, erect, massive and boring morphologies (Fig. 7.4), all influenced by ambient environmental conditions such as current flow and substrate. Any given sponge species may assume a variety of growth forms depending upon the prevailing environmental conditions. The pattern of growth may be influenced by the nature and inclination of the substrate, available space, and the velocity and type of water currents.

The spicules form the hard sponge skeleton. Spicules are classified in two broad groups: megascleres, which are the larger spicules that form the main supporting elements of the skeleton; and smaller spicules called microscleres. The fossil record of sponges includes

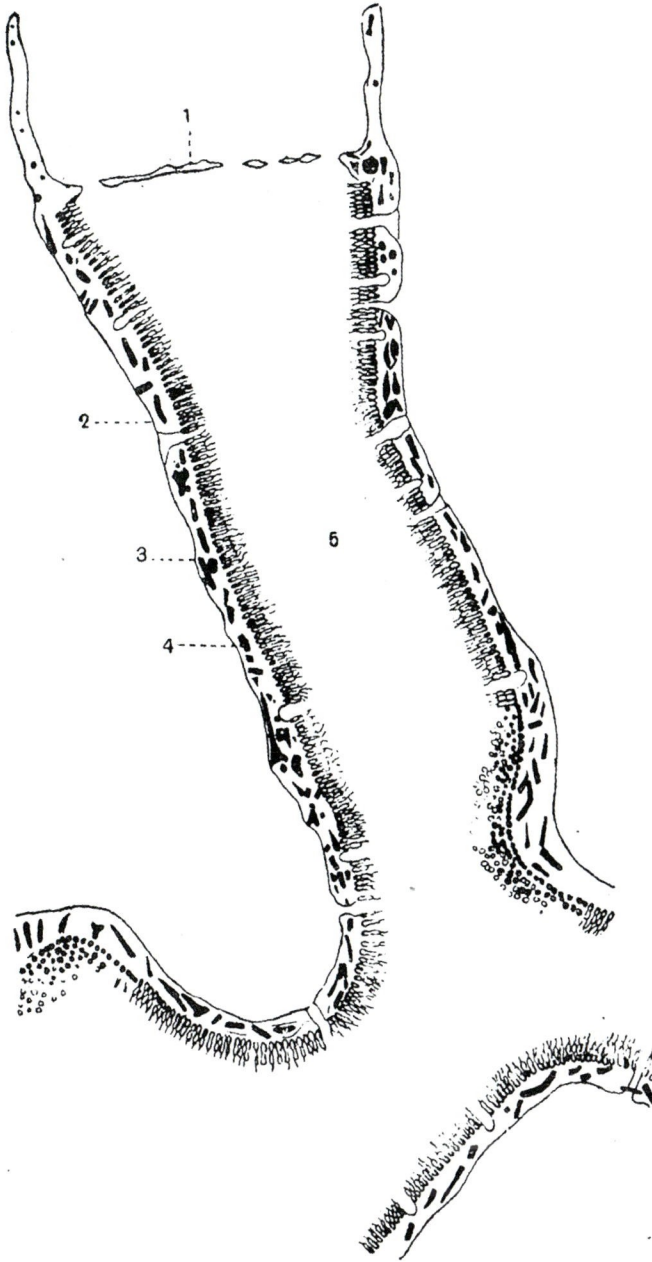

**Figure 8.3** Highly
magnified vertical
section through part of
the calcareous sponge
*Leucosalenia* sp. (after
Woods, 1955, fig. 7). Key:
1 = sieve-like membrane
covering the osculum; 2
= outer layer; 3 = collar
or flagellated cells; 4 =
spicules; 5 = gastral cavity
(spongocoel).

isolated spicules, spicule mats and partial or complete sponges. Spicules are of taxonomic importance and fall into various morphologic groups (Clarkson, 1998, chapter 4).

In the geologic past, sponges have sometimes been important reef-building organisms and have been a source of biogenic silica (chert bands in limestone sequences are often rich in sponges). Boring clionaid sponges were and are important bioeroders. Some sponge-rich horizons have been used as marker bands in biostratigraphy.

## On the beach

The seashores of the British Isles are not a mecca of modern stony corals and sponges except where they might be washed ashore in major storms. Fortunately, if you intend to travel abroad in search of stony corals, more equatorial beaches will likely be littered with storm debris reworked from reefs offshore. To mention one example that I know well, modern Caribbean coral reefs typically have a diverse coral fauna. For example, 62 species are known from the Discovery Bay and Runaway Bay region of central north Jamaica (Huston, 1985a, b). The reef coral fauna shows a depth-related zonation. As depth increases, wave energy decreases, sunlight filtering through the water column decreases and temperature decreases. Different coral species, adapted to distinct suites of physical conditions, are found at different depths. The greatest diversity of taxa is found at the shallowest depths – for example, about 50 species in the top 30 m at Discovery and Runaway Bays. The growth rate of a shallow water coral species commonly decreases with increasing depth.

Evidence for one group of sponges, the clionaids, is provided by their borings, common in limestones, mudrocks and shells. This trace, *Entobia* isp., is a regular feature of many beaches in the British Isles and elsewhere (Figs 7.4, 17.4B–D, 18.5C, E, amongst others; Donovan, 2021, figs 48.5B, C, 53.3, amongst others).

## Guides to identification

Stony corals are not a common or diverse part of the marine fauna around the British Isles. Adequate notes for identification can be found in general guides like Barrett & Yonge (1977, pp. 60–61) and Campbell (1982, pp. 98–99). But if stony corals are a group of particular interest to you and you intend to go further afield, there are certainly numerous guides to local marine biotas in the humid tropics as well as many online reference sites. Your problem will be linking books for identification of live stony corals, perhaps aimed at the diver, with the clasts of dead coral skeleton found on the shore. Fortunately, these clasts can be collected like pebbles on the seashore and identified at your leisure.

## References

Barrett, J. & Yonge, C.M. (1977) (first published 1958) *Collins Pocket Guide to the Sea Shore*. Collins, London.
Benton, M.J. & Harper, D.A.T. (1997) *Basic Palaeontology*. Addison Wesley Longman, Harlow.
Campbell, A.C. (1982) (first published 1976) *The Country Life Guide to the Seashore and Shallow Seas of Britain and Europe*. Country Life Books, London.
Clarkson, E.N.K. (1998) *Invertebrate Palaeontology and Evolution*. 4th ed. Blackwell Science, Oxford.
Donovan, S.K. (2021) *Hands-On Palaeontology: A Practical Manual*. Dunedin Academic Press, Edinburgh.
———. (2023) Three views: Complex post-exhumation history of a Chalk cobble, north Norfolk. *Bulletin of the Geological Society of Norfolk*, **73**: 85–93.
Huston, M.A. (1985a) Variation in coral growth rates with depth at Discovery Bay, Jamaica. *Coral Reefs*, **4**: 19–25.
———. (1985b) Patterns of species diversity on coral reefs. *Annual Review of Ecology and Systematics*, **16**: 149–177.
Oliver, W.A. Jr. & Coates, A.G. (1987) Phylum Cnidaria. In Boardman, R.S., Cheetham, A.H. & Rowell, A.J. (eds), *Fossil Invertebrates*: 140–193. Blackwell Scientific, Palo Alto, CA.
Woods, H. (1955) *Palaeontology: Invertebrate*. 8th ed. Cambridge University Press, Cambridge.

# CHAPTER 9

# Bryozoans

## Form and function

The bryozoans are colonial organisms which possess a lophophore and are thus related to brachiopods (Clarkson, 1998). Although bryozoans are superficially similar to colonial corals, they have a free suspended gut with both a mouth and an anus, whereas cnidarians have a single opening to a blind gut. All bryozoans are aquatic and most are marine. The colony is constructed by the tiny bryozoan animals, called zooids (Figs 9.1, 9.2), which are commonly partially enclosed in a calcitic tube into which they may be withdrawn for protection. The zooid tube is called a zooecium (Fig. 9.2A) if calcified and a cystid if non-calcified. The whole calcified colony is called the zoarium. Although generally small, the largest known bryozoan colonies are about 1 m in diameter. The zooids are cylindrical. The lophophore has a ciliated crown of tentacles (Figs 9.1A, 9.3). These tentacles are adapted for filter feeding and radiate from a central mouth. The U-shaped gut (Fig. 9.1A) connects the mouth to the anus, which opens outside the ring of tentacles. The tentacles are extended for feeding or retracted for protection. Retraction is by the retractor muscle, while eversion is produced by the transverse muscles compressing the body wall. All zooids are connected by a funicular tube that attaches to the base of the gut by a funiculus thread (Fig. 9.1A). Specialised zooids are adapted for brooding larvae or defence.

The initial zooid is called the ancestrula. New zooids are budded from the ancestrula asexually, while new colonies are produced sexually. Fertilised eggs develop into free-swimming larvae, while sperm is also released to fertilise other colonies. In cheilostomes (see below) the fertilised egg may develop into a larva in an ovicell or ooecium (Fig. 9.2A), a structure developed in mature zooids.

The orifice or opening of the zooid is covered by an operculum in cheilostomes. The upper surface, called the frontal wall, is comprised of several layers and is often complexly sculptured (Figs 9.2A, 9.3). Cheilostome zooids are polymorphic. For example, in avicurala zooids (Fig. 9.1B, 9.2A) the retractile portion of the zooid (called the polypide) is modified as a snapping structure that looks like the head of a bird and discourages larval settlement and predation. Similarly, vibraculum zooids are lashing bristles which discourage larval settlement.

The bryozoan colony can take many growth forms, including branching and sheet-like encrustations (Fig. 9.4); branched, frond-like and domed erect colonies (Figs 9.5, 9.6);

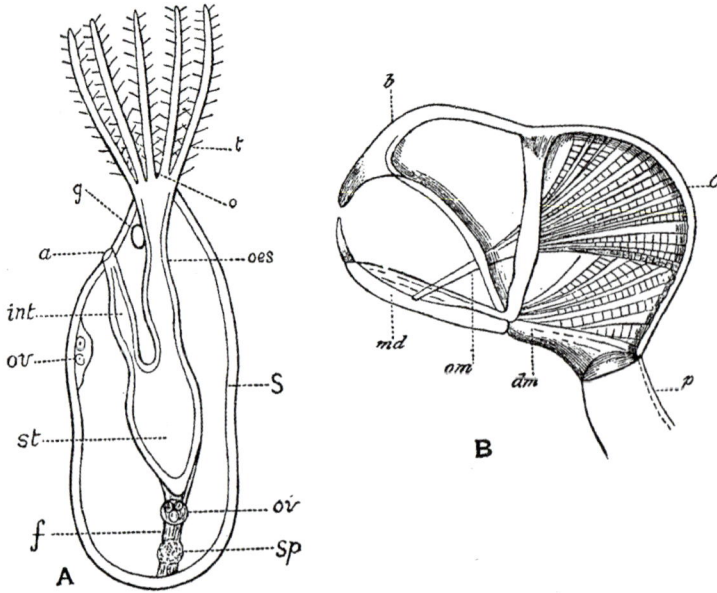

**Figure 9.1** (After Woods, 1955, fig. 120.) (**A**) The structure of a single bryozoan individual (= zooid). Key: S = body-wall; t = tentacles; o = mouth; oes = oesophagus; st = stomach; int = intestine; a = anus; g = ganglion; f = funiculus; ov = ovary; sp = testis. Enlarged. (**B**) Avicularium of cheilostome *Bugula* sp. Key: b = beak; md = mandible; C = chamber; p = peduncle; om = occlusor muscles; dm = divaricator muscles. Enlarged.

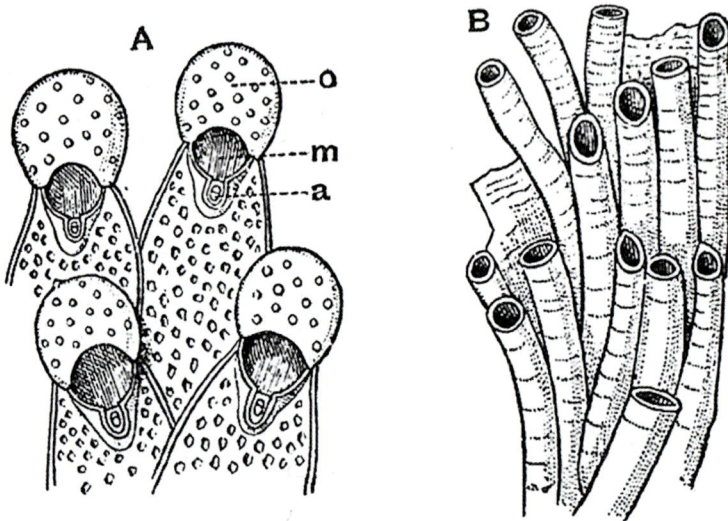

**Figure 9.2** (After Woods, 1955, fig. 121.) Details of extant bryozoans, enlarged. (**A**) Portion of *Smittia landsborovi* (Johnston), a cheilostome. Key: o = ooecium; m = aperture of the zooecium; a = avicularium. (**B**) Portion of *Tubulipora fimbria* Lamarck, a cyclostome.

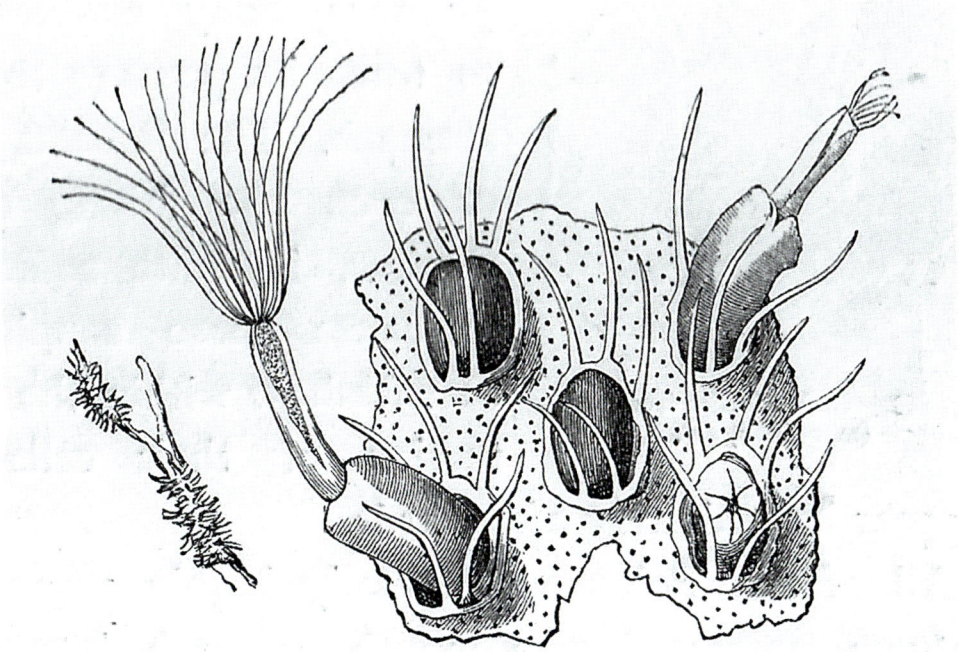

**Figure 9.3** (After Taylor, 1885, fig. 167.) Recent bryozoan, showing polyps protruded (magnified). Compare with Figures 9.1A and 9.2A.

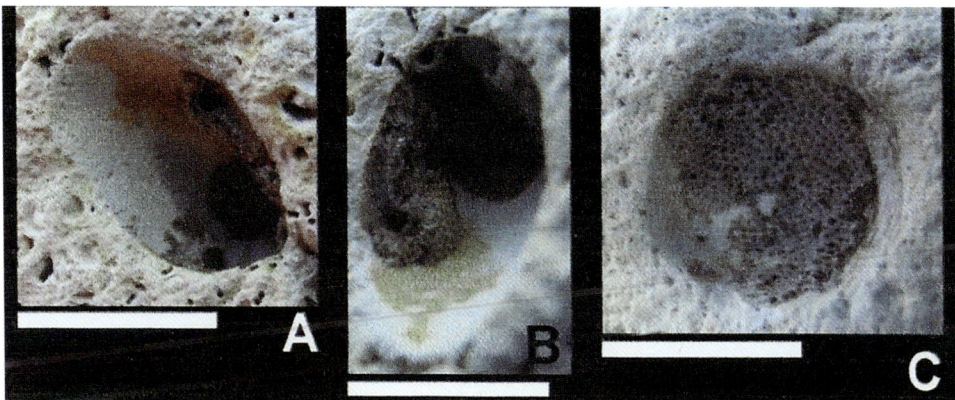

**Figure 9.4** Chalk cobbles from the beach between Overstrand and Cromer, north coast of Norfolk, England (after Donovan, 2017, fig. 3A–C). Specimens are deposited in the Naturalis Biodiversity Center, Leiden, the Netherlands (prefix RGM). (**A, B**) RGM 792 405, two views of the base and part of the chamber of the bivalve boring *Gastrochaenolites* isp. encrusted by agglutinated worm tubes, which were, in turn, encrusted by the cheilostome bryozoan *Cryptosula pallasiana* (Moll). (**C**) RGM 792 406, encrusting *C. pallasiana* in *Gastrochaenolites* boring. Scale bars represent 10 mm.

**Figure 9.5** Sea mat or hornwrack (*Flustra foliacea* (Linné)) (after Anon., 1938, p. 15, upper left), a seaweed-like bryozoan. Recent. The inset shows the zooids in life.

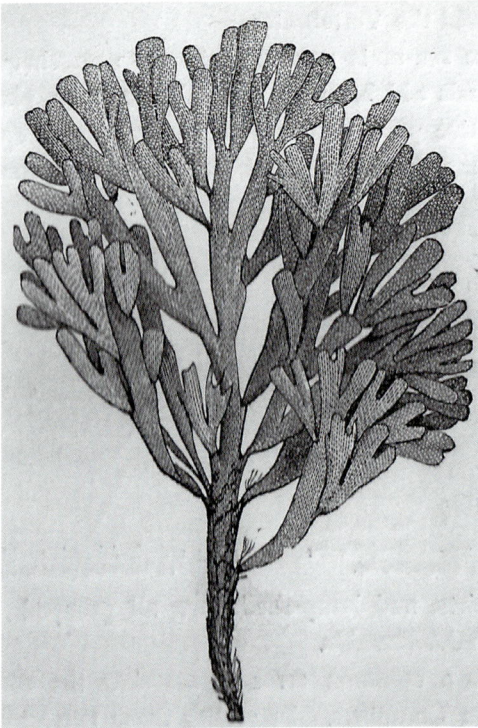

**Figure 9.6** (After Taylor, 1885, fig. 167.) Recent common sea-mat, *Flustra* sp. Maximum dimension *c*. 90 mm.

and borers into calcium carbonate substrates. Bryozoan colonies are mainly immobile, although the extant lunulitiform bryozoans have evolved specialised walking vibraculum zooids. Encrusters are common on the shells of living and dead organisms, and on hard, inorganic substrates. Erect colonies (= 'stick' bryozoans) favoured shallow-water marine, carbonate-rich environments and are generally preserved as twig-like fragments. The zooids of the frond or vase-like colony of Lower Carboniferous *Fenestella* sp. (Donovan & Wyse Jackson, 2018; Fig. 9.7 herein) may have generated feeding currents throughout the colony.

From 15,000 to 20,000 fossil species of bryozoan are known, plus at least 3,500 that are extant, divided between seven orders. The bryozoans, now known from the Cambrian (Zhang *et al.*, 2021), underwent an abrupt radiation in the Early Ordovician and were drastically depleted in diversity by the end Permian mass extinction event. A major radiation of the order Cheilostomata occurred in the Late Cretaceous. The cheilostomes are the dominant extant order, the zooid having evolved many complex structures. This radiation of the cheilostomes may have been related to the evolution of a new type of larva. Primitive 'malacostegan' cheilostomes have a planktotrophic (feeding) larva, while advanced cheilostomes have a non-planktotrophic larva. Gene flow is poorer in taxa with a non-planktotrophic larva, which may have encouraged speciation. Larval brooding in the cheilostomes (Fig. 9.2A, ooecium) first appears in the fossil record immediately prior to this radiation event. Different groups have shown repeated convergent evolution of similar colony forms, such as reticulate colonies evolved in *Fenestella* in the Upper Palaeozoic and in two other orders in the post-Palaeozoic.

## On the beach

Bryozoans on British beaches can be divided into two preservational groups, those that have a fossil-like occurrence and the rest. Obviously, it is the first cluster that is of most interest to us and it is composed of bryozoans with a calcareous shell, mainly cheilostomes, which encrust hard substrates such as shells and (commonly bored) cobbles. I emphasise the association between limestones and bryozoans which secondarily infest discarded borings (Fig. 9.4). Common invaders that cement within discarded borings such as *Gastrochaenolites* isp. include bryozoans, serpulids and spirorbids (Donovan, 2017; see Chapter 15 herein).

**Figure 9.7** (After Donovan & Wyse Jackson, 2018, fig. 2.) A fenestrate bryozoan colony (Mississippian) seen in a building stone, but similar occurrences may be common in limestone beach cobbles. (**a**) Complete specimen showing simple branching pattern of a vase-shaped colony exposed in two dimensions. Scale in cm. (**b**) Detail of part of the structure showing 'windows' in structure (= fenestrae); zooids occur in the calcareous 'frames' to the fenestrae. Scale bar represents 10 mm.

Preservation on hard substrates can be selective. On shells, the occurrence of encrusters is most commonly found in protected areas such as on the inner, concave surfaces of valves and within the aperture of gastropods. Such infestations obviously occur after the death of the mollusc. They are areas protected from abrasion during transport on the seafloor. Some bryozoans may encrust the convex outer surfaces of shells, but here they are likely to be ground off during transport. Evidence of this can sometimes be found on convex external surfaces which are strongly ribbed. Careful examination may reveal that part of the colony is preserved in the protected grooves between ribs, having been worn away on the crests of the ribs.

A common seaweed-like, brown bryozoan on British beaches is the unmineralised cheilostome *Flustra* (Figs 3.1, 9.5, 9.6), commonly known as hornwrack or sea-mat. Although unlikely to be fossilised, lacking a calcified skeleton, *Flustra* may be washed ashore in great numbers and is worthy of note. Torn up during storms from its attachments to the substrate offshore, these colonies may be large, up to 15 cm maximum dimension.

Also notice chalk and limestones cobbles which may preserve fossil bryozoans, usually in section. That is, these three-dimensional structures are apparent in two dimensions, making them harder to recognise and identify. The figured example (Fig. 9.7) is from the wall of a bank in the Netherlands (Donovan & Wyse Jackson, 2018) but might equally have been found in a clast of Carboniferous limestone on the beach.

### Guides to identification

Readable and succinct introductions to the bryozoans are provided by Clarkson (1998, pp. 143–157) and Ryland (1970). They receive only short shrift in standard field guides (e.g., Kosch *et al.*, 1963, pp. 60–61; Barrett & Yonge, 1977, pp. 171–174; Campbell, 1982, pp. 232–237), although they are moderately diverse (Bruce *et al.*, 1963, pp. 225–238). The series 'Synopses of the British Fauna' includes several volumes on marine and freshwater bryozoans (e.g., Hayward & Ryland, 1977).

### References

Anon. (1938) *The Sea-Shore*. W.D. & H.O. Wills, Bristol.

Barrett, J. & Yonge, C.M. (1977) (first published 1958) *Collins Pocket Guide to the Sea Shore*. Collins, London.

Bruce, J.R., Colman, J.S. & Jones, N.S. (1963) *Marine Fauna of the Isle of Man*. Liverpool University Press, Liverpool.

Campbell, A.C. (1982) (first published 1976) *The Country Life Guide to the Seashore and Shallow Seas of Britain and Europe*. Country Life Books, London.

Clarkson, E.N.K. (1998) *Invertebrate Palaeontology and Evolution*. 4th ed. Blackwell Science, Oxford.

Donovan, S.K. (2017) Neoichnology of Chalk cobbles from north Norfolk, England: Implications for taphonomy and palaeoecology. *Proceedings of the Geologists' Association*, **128**: 558–563.

Donovan, S.K. & Wyse Jackson, P.N. (2018) Well-preserved fenestrate bryozoans in Mississippian building stones, Utrecht, The Netherlands. *Swiss Journal of Palaeontology*, **137**: 99–102.

Hayward, P.J. & Ryland, J.S. (1977) *British Ascophoran Bryozoans*. Linnean Society/Academic Press, London.

Kosch, A., Frieling, H. & Janus, H. (1963) *The Young Specialist Looks at Seashore*. Burke, London.

Ryland, J.S. (1970) *Bryozoans*. Hutchinson, London.

Taylor, J.E. (1885) *Our Common British Fossils and Where to Find Them*. Chatto & Windus, London.

Woods, H. (1955) *Palaeontology: Invertebrate*. 8th ed. Cambridge University Press, Cambridge.

Zhang, Zhil., Zhang, Zhif., Ma, J., Taylor, P.D. *et al.* (2021) Fossil evidence unveils an early Cambrian origin for Bryozoa. *Nature*, **599**: 251–255.

# CHAPTER 10

# Crabs

## Form and function

I choose to concentrate this chapter mainly on brachyuran crabs, but not shrimps or lobsters, for several valid reasons. Crabs are common on British beaches, have durable carapaces and chelae (= 'claws') (Fig. 10.1), may be present as monospecific accumulations of exuviae at times of moulting and have a good fossil record. Lobsters are relatively rare, and shrimps are less robust, although they may be found alive, with crabs, in rock pools. If shrimps and lobsters are a particular interest, then refer to Smaldon *et al.* (1993) and Ingle & Christiansen (2004). The one addition to the brachyurans that will be mentioned herein are the anomuran or hermit crabs (Fig. 10.2).

The decapod crustaceans (Devonian to Recent) include the crabs, lobsters, crayfishes and shrimps. The name decapod is derived from the five pairs of thoracic appendages that are used mainly for walking, but which show a range of modifications. In particular, the anterior pair of thoracic legs are adapted as claws called chelae (Figs 10.1, 10.3), which are capable of tasks such as crushing shells and cutting food. The chelae are usually well-calcified and

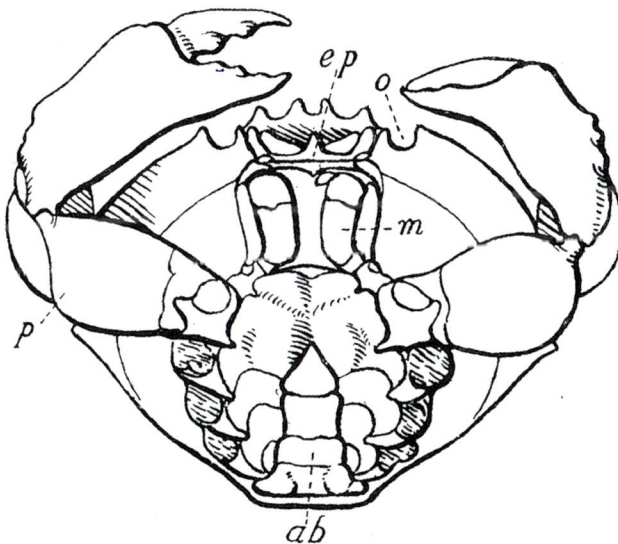

**Figure 10.1** *Xanthopsis dufourii* (Milne-Edwards), Eocene (after Woods, 1955, fig. 208). Ventral view. Key: ep = epistome (mouth region); o = orbit (for the eye); m = third maxillipede (= 'jaw leg' for passing food to the chewing mouthparts of the head proper); p = first thoracic leg (cheliped or chela); ab = abdomen (folded under cephalothorax).

**Figure 10.2** Hermit crab (*Pagurus bernhardus* (Linné)) in shell of a whelk, *Buccinum* sp. (after Wood, 1905, fig. on p. 165).

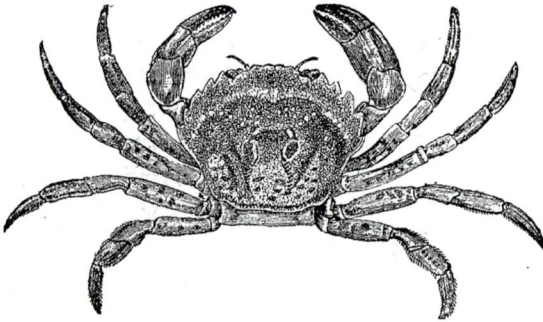

**Figure 10.3** Common shore crab, *Carcinus maenas* (Linné) (after Wood, 1905, fig. on p. 161). Dorsal view. Note ten appendages; the chelae are anterior. Compare with Figure 10.4.

therefore have a good fossil record, especially amongst the crabs; modern examples may be common on beaches. Preservation of 'complete' decapods is variable, because the carapace is not always well-calcified, and the skeleton is comprised of multiple elements which disarticulate after death. Because it adds to the weight of the exoskeleton, calcification is usually better developed in walking, rather than swimming, decapods. Calcite is resorbed during ecdysis, the process of shedding the outer cuticle, a characteristic of arthropods.

The decapod body is divided into an anterior cephalothorax, the fused head plus thorax, protected dorsally and ventro-laterally by the carapace (e.g., Figs 10.1, 10.3), and a posterior abdomen (Fig. 10.1); in crabs, the latter is folded under the ventral surface of cephalothorax. In normal locomotion, the thoracic legs are used, but a defensive reaction is to swim backwards by flexing the abdomen under the cephalothorax, a movement aided by the flattening and expansion of the last abdominal segment as a paddle-like telson. Predators must therefore approach such a fleeing decapod head-on and face the paired chelae and often spinose anterior. Predators may be further discouraged by decapods which camouflage their exoskeleton with algae and sea anemones.

An extreme adaptation amongst decapods is shown by hermit crabs (family Paguridae; Fig. 10.2). Hermit crabs use the discarded shells of gastropods as dwellings. One chela may be larger and adapted as an 'operculum' to seal the aperture once the crab has withdrawn into the shell. Hermit crabs sometimes modify the aperture of the shell to their own requirements by breaking off fragments with their chelae. Unlike gastropods, the shell is dragged along the substrate, rather than held above it, so that the shell may become abraded. Recognition of these features in fossil gastropod shells has sometimes been used as evidence of ancient 'pagurisation'.

## On the beach

Live crabs may be common in rock pools and on rocky shorelines. While all hard parts of dead crabs may be found, it is the carapaces and chelae that are most easily identifiable and collectible.

The life habit of a crab may be determined from the evidence of carapace thickness, and the carapace shape and its protective adaptations (Schäfer, 1972, pp. 133–137). Schäfer

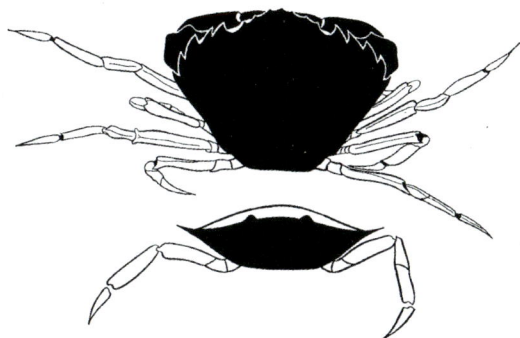

**Figure 10.4** *Carcinides maenas* (Linné) (after Schäfer, 1972, fig. 79), a crawling/running brachyuran. **Top**. Dorsal view with chelae folded against the carapace and other elongate, walking legs extended. **Bottom**. Cross-section. Maximum dimension about 150 mm wide over tips of legs. Compare with Figure 10.3.

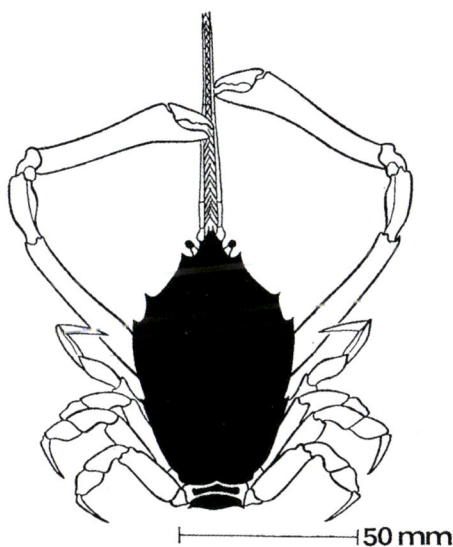

**Figure 10.5** *Corystes cassivelaunus* (Pennant), a digging/burrowing species, dorsal view (after Schäfer, 1972, fig. 80). Note the long, parallel antennae with a 'snorkel' formed by hairs. *Corystes* reverses in to its burrow, leaving the snorkel exposed.

50 mm

**Figure 10.6** Climbing (spider) crabs. **Left**: *Hyas araneus* (Linné); **right**: *Macropodia rostrata* (Linné) (after Schäfer, 1972, fig. 81). Dorsal views. Both to scale, length of left carapace about 90 mm.

**Figure 10.7** The swimming crab *Liocarcinus holsatus* (Fabricius) (after Schäfer, 1972, fig. 82 *pars*). Swimming, dorsal view; left appendages incomplete. Carapace about 50 mm wide.

divided crabs into four functional groups based on habit reflected by morphology: crawlers and runners (Fig. 10.4); diggers (Fig. 10.5); climbers (Fig. 10.6); and swimmers (Fig. 10.7). Crawlers and runners are the most likely crabs to be encountered live on a beach. The body is compact; a lack of protrusions offers little resistance to strong currents when in shallow

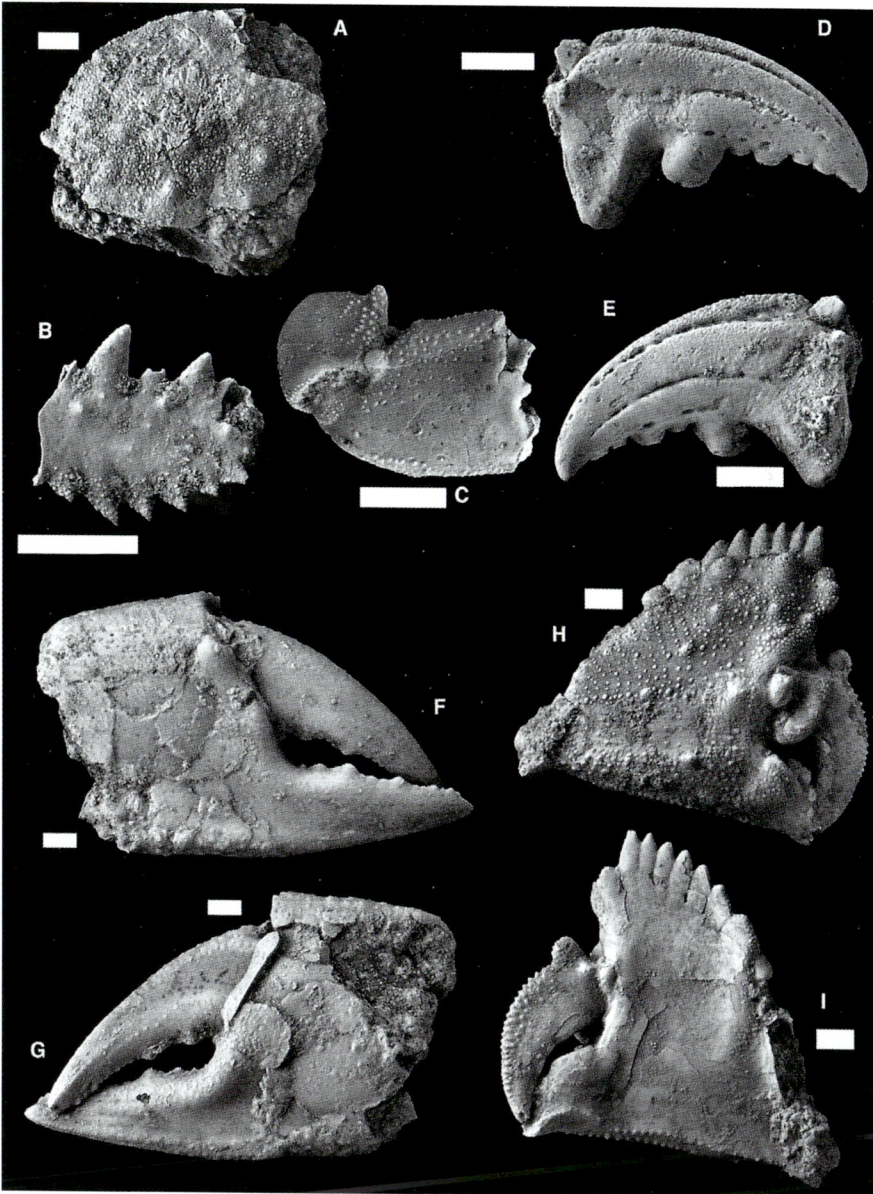

**Figure 10.8** Mainly chelae and other fragments (after Donovan & Collins, 2013, fig. 5). Fossil decapod crustaceans from the Late Pleistocene Port Morant Formation, south-east Jamaica. **A, H, I**. *Calappa springeri* Rathbun. **A**. Fragment of carapace. **H, I**. Right chela, outer (**H**) and inner surfaces. **B**. *Platylambrus* sp. cf. *P. serratus* H. Milne Edwards, distal end of left propodus (= the immovable finger that extends up from the palm of a chela). **C**. *Eurypanopeus* sp. cf. *E. depressus* (Smith), right propodus and associated carpus (= joints of the legs). **D, E**. *Panopeus rugosus* A. Milne Edwards, right dactylus (= free finger), outer (**D**) and inner surfaces. **F, G**. *Cardisoma guanhumi* Latreille, right chela, outer (**F**) and inner surfaces. All specimens coated with ammonium chloride for photography. Scale bars represent 4 mm.

water. The body is flattened dorso-ventrally (Fig. 10.4, lower) with the chelae fitting into recesses in the carapace. The walking legs (pereiopods) are long and strong, well-adapted for locomotion. The illustrated example, *Carcinides maenas* (Linné) (Figs 10.3, 10.4), is a typically fast-moving species with a light and thin carapace.

Diggers (Fig. 10.5) have an elongate carapace that narrows posteriorly. The anterior spines point forward, and the cephalothorax is smooth, facilitating burrowing. The antennae may be long, stiff and parallel, forming a breathing tube. (I collected a specimen with a particularly well-preserved breathing tube when I taught in the West Indies; between one year and the next, all of the 'hairs' between antennae were eaten by insects!) The locomotory limbs decrease in size posteriorly; the chelae are too big to fold against the carapace. The most posterior limbs are adapted for digging.

Climbers include the spider crabs which do resemble spiders (Fig. 10.6). The cephalo-thorax is pear-shaped and narrows anteriorly (contrast Figs 10.5 and 10.6). The carapace is rough and thorny, improving adhesion to all substrates. All limbs are long and spider-like with long claws. Climbers are the only one of the four morphological groups considered herein that cannot burrow.

Swimmers (Fig. 10.7) have a gross morphology reminiscent of walkers. The main difference is the locomotory limbs are flattened, particularly posteriorly, where they are paddle-like. The carapace is thin, presumably an adaptation to reduce weight. Again, one of my West Indian experiences is of relevance, when I was snorkelling and disturbed a swimming crab; it whirled away violently like some kind of floating clockwork toy.

Fragments of limbs, particularly the chelae (Fig. 10.8), may be common on any beach. In the fossil record, the majority of crab fossils are preserved as limb fragments, of which chelae are the biggest and most obvious. These can be identified to species, with care.

## Guides to identification

The master reference on British crabs is Ingle (1980). I admit to giving my copy away when I retired and had to buy a second copy afterwards because I realised that I couldn't live without it! Other standard references to the seashore, listed in many chapters of this section, will also be useful. For hermit crabs, see Ingle & Christiansen (2004).

## References

Donovan, S.K. & Collins, J.S.H. (2013) Jamaican fossil crabs. *Deposits*, **33**: 13–19.

Ingle, R.W. (1980) *British Crabs*. British Museum (Natural History)/Oxford University Press, Oxford.

Ingle, R.W. & Christiansen, M.E. (2004) *Lobsters, Mud Shrimps and Anomuran Crabs*. Field Studies Council Publications, Shrewsbury.

Schäfer, W. (Craig, G.Y., ed.) (1972) *Ecology and Palaeoecology of Marine Environments*. University of Chicago Press, Chicago.

Smaldon, G., Holthuis, L.B. & Fransen, C.H.J.M. (1993) *Coastal Shrimps and Prawns*. Field Studies Council Publications, Shrewsbury.

Wood, J.G. (1905) *The Common Objects of the Sea-Shore Including Hints for an Aquarium*. 16th ed. George Routledge & Sons, London.

Woods, H. (1955) *Palaeontology: Invertebrate*. Cambridge University Press, Cambridge.

# CHAPTER 11

# Barnacles

The following notes are based on Collins *et al.* (2014, pp. 216–217). The barnacles, or Cirripedia, form a significant part of the arthropod assemblages of beaches of the British Isles. Whilst certain species of balanid (= acorn barnacles) are readily distinguishable, others may easily be confused, particularly those that occur only as disarticulated plates. Specific determination may commonly be achieved by the recognition of either any one, or a combination, of a few major features. The general appearance of the balanid shell, whether smooth or ribbed, conical or cylindrical, and features such as the shape of the orifice and nature of the radii and alae, may, with practice, be sufficient evidence on which to base a specimen's identification (for explanation of morphological terminology, see Fig. 11.1). Nevertheless, these features should not be relied upon entirely, since the appearance of the shell varies, as observations of living forms show, according to habitat. For example, cramped conditions give rise to distorted forms, whilst exposed individuals commonly develop thickened shells; the nature of the substrate may also influence form of growth. The nature of the opercular valves (terga and scuta; Fig. 11.1F–M) of balanids, when preserved, is generally sufficient to provide tolerable determination. They are seldom preserved *in situ* in fossil and recently dead individuals, but when these valves are not obvious, it is always sound practice to probe carefully into the orifice of the shell to see if they are among the infilling debris. However, in many cases the valves are lost completely, and it is necessary to resort to an examination of the shell. The compartments (walls or parietes), basis and radii of the shell are sometimes composed of inner and outer laminae which are joined together by septa, forming a series of tubes or pores. The presence or absence of these pores and the construction of the septa provides a useful, albeit not always final, indication of the species. The relative angle of the ala and radius to the paries (plural, parietes) is another important consideration. The balanids attach to the substrate by a basal plate, or basis; if an individual is lost from a substrate, it may still leave behind an attachment scar as evidence (Miller & Brown, 1979)

## Form and function

**Goose barnacles**: The goose, stalked or pedunculate barnacles range from the Carboniferous to Recent. Goose barnacles (Figs 7.1, 11.2) differ markedly from acorn barnacles in having a shell – the capitulum – that is laterally compressed, in being stalked (the peduncle) and, with one exception around the British Isles (*Pollicipes pollicipes* (Gmelin); Southward, 2008),

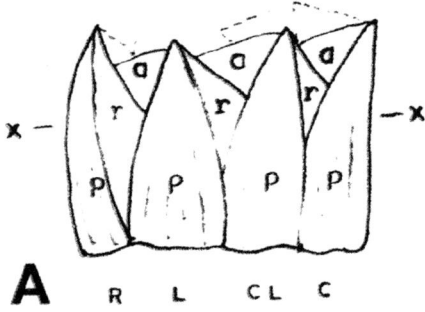

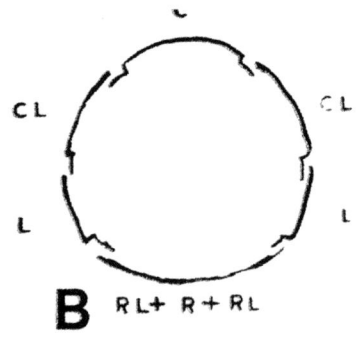

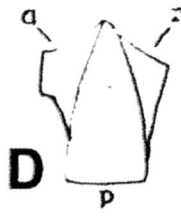

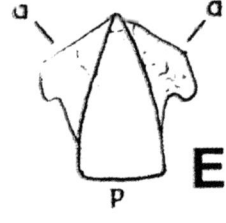

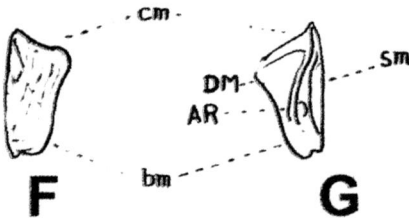

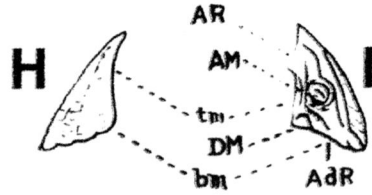

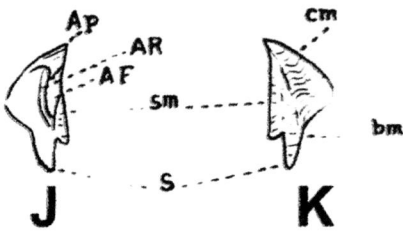

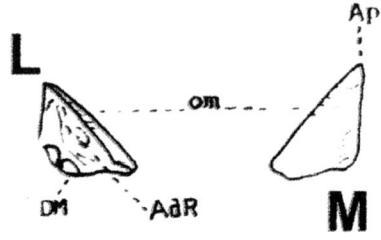

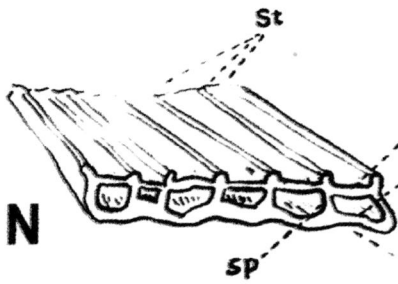

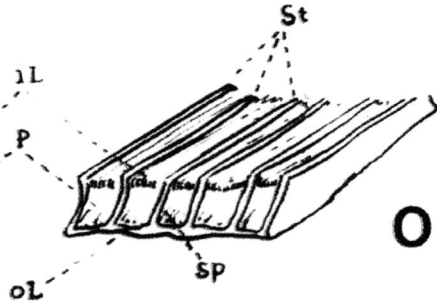

**Figure 11.1** Features of the shell of the balanid barnacles (after Collins *et al.*, 2014, fig. 1). (**A, B**) Gross morphology of *Balanus*. (**A**) Lateral view; (**B**) section across the shell X–X (see **A**) showing the mode of imbrication of the compartments (six in this genus). The rostrum and rostrolateral compartments (RL) are combined; it is customary to refer to this composite compartment simply as the 'rostrum'. Key: a = alae; p = parietes; r = radii; C = carina; CL = carinolateral; L = lateral; R = rostrum; RL = rostro-lateral. (**C**) Compartment with two radii, serving either as a rostrolateral or as a rostrum combined with the rostrolaterals, as in *Balanus*. (**D**) Compartment with a radius and anala, serving as a lateral or carinolateral compartment. (**E**) Compartment with two alae, serving as a carina (also as a rostrum in genera other than *Balanus*). (**F–I**) External (**F, H**) and interior views (**G, I**) of terga of various forms; (**J–M**) external (**K, M**) and interior views (**J, L**) of scuta of various forms. Key: bm = basal margin; cm = carinal margin; om = opercular margin; sm = scutal margin; tm = tergal margin; AdR = adductor ridge; AF = articular furrow; AM = (circular) pit for adductor muscle; Ap = apex; AR = articular ridge; DM = pit for lateral depressor muscle; S = spur. (**N, O**) Sections, magnified many times, through parietes to show the arrangement of the septa and pores in two examples of shell construction, typical of *Balanus crenatus* Brugière (**N**) and *Balanus balanus* Linné (**O**).

**Figure 11.2** Coiled shell of the deep-sea cephalopod *Spirula spirula* (Linné) encrusted by the goose barnacle *Lepas anatifera* Linné, the Natural History Museum, London, Zoology Department specimen 1988•373 (after Donovan, 1989b, fig. 1). The barnacle was transported as pseudoplankton but could only attach to *Spirula* after the shell was released by the rotting soft tissues after death. Scale bar represents 5 mm. See also Figure 14.1.

are not attached intertidally. The shell, like that of balanids, is composed of multiple plates that are less well sutured together and disarticulate soon after death.

Goose barnacles are commonly found attached to floating debris that has been beached, such as wood, cephalopod shells and plastic bottles (Fig. 11.2). Such occurrences of non-planktonic organisms attached to floating objects is termed pseudoplankton (Wignall & Simms, 1990).

**Acorn barnacles**: The Balanomorpha, or acorn barnacles (Figs 11.1, 11.3, 11.4), range from the Upper Cretaceous to Recent. The acorn barnacles are a particularly successful group, having invaded the intertidal zone on rocky shorelines. It is this group which led Darwin to name the Holocene 'The Age of Barnacles'. Acorn barnacles lack a peduncle and cement directly to the substrate, with the secretion of either a calcareous basal plate or a membraneous basis over the hard substrate. The capitulum (shell) is conical and composed of four to eight overlapping compartmental plates, which are more rigidly articulated than the shell in goose barnacles. The compartmental plates have a distinctive internal arrangement of longitudinal tubes. This open network permits rapid growth and accounts for the success of the balanids when compared with other, slower-growing barnacle groups. The aperture is broad and sealed by four cover plates. Members of this group are almost exclusively hermaphroditic.

## On the beach

My interests in extant and Neogene barnacles are mainly concerned with their ecology and palaeoecology rather than their classification (such as in Donovan, 1988, 1989a). Barnacles live intimately attached to their substrate. Pedunculate barnacles are more likely to be found attached to floating flotsam washed up on a beach, that is, as pseudoplankton. Two examples are demonstrated herein. Of particular interest to the palaeontologist are barnacles attached to floating cephalopod shells and any evidence this might provide for our studies of extinct taxa such as ammonites (Donovan, 1989b). The coiled conch of *Spirula spirula* (Linné), encrusted by the goose barnacle *Lepas anatifera* Linné (Figs 11.2, 14.1), is reminiscent of an ammonite, but in life the shell was internal (Fig. 14.1). It can only be encrusted after death of *Spirula* and release of the shell from the rotting body.

Another geologically relevant specimen is the attachment of multiple *L. anatifera* to a cobble of pumice (Fig. 7.1). The pumice probably erupted from the Soufriere Hills of Montserrat in the late 1990s, but was collected on the Palisadoes, Jamaica (Donovan, 2025), about 1,500 km away (Donovan, 1999).

Balanids may be found on other shells and rocky substrates. *Ensis americanus* (Binney) is an alien razor shell living infaunally in the southern North Sea. It is encrusted postmortem by the acorn barnacle *Balanus crenatus* Brugière. Remarkable shells of *E. americanus* are found densely infested by one or more spatfalls of *B. crenatus* on the inner and outer surfaces of both valves. The ligament between the valves is still attached and may be pliant when collected, indicating that infestation was soon after death of the bivalve, probably only weeks or months before it was collected. The evidence of these specimens argues strongly against the common received wisdom that borings or encrustations on the inner surfaces of disarticulated fossil bivalves are an indication of a relatively long postmortem residence time on the seafloor (Fig. 11.3; Donovan, 2007, 2011; Donovan *et al.*, 2014)

Finally, there is one interesting Neogene fossil example (Fig. 11.4; Donovan & Novak, 2015). Site selectivity of predatory boreholes is a well-known phenomenon in modern and ancient shells (Pickerill & Donovan, 1998). A broken oyster valve from the Upper Pliocene of Rambla de la Sepultura in the Almeria-Nijar Basin, south-east Spain, is encrusted by

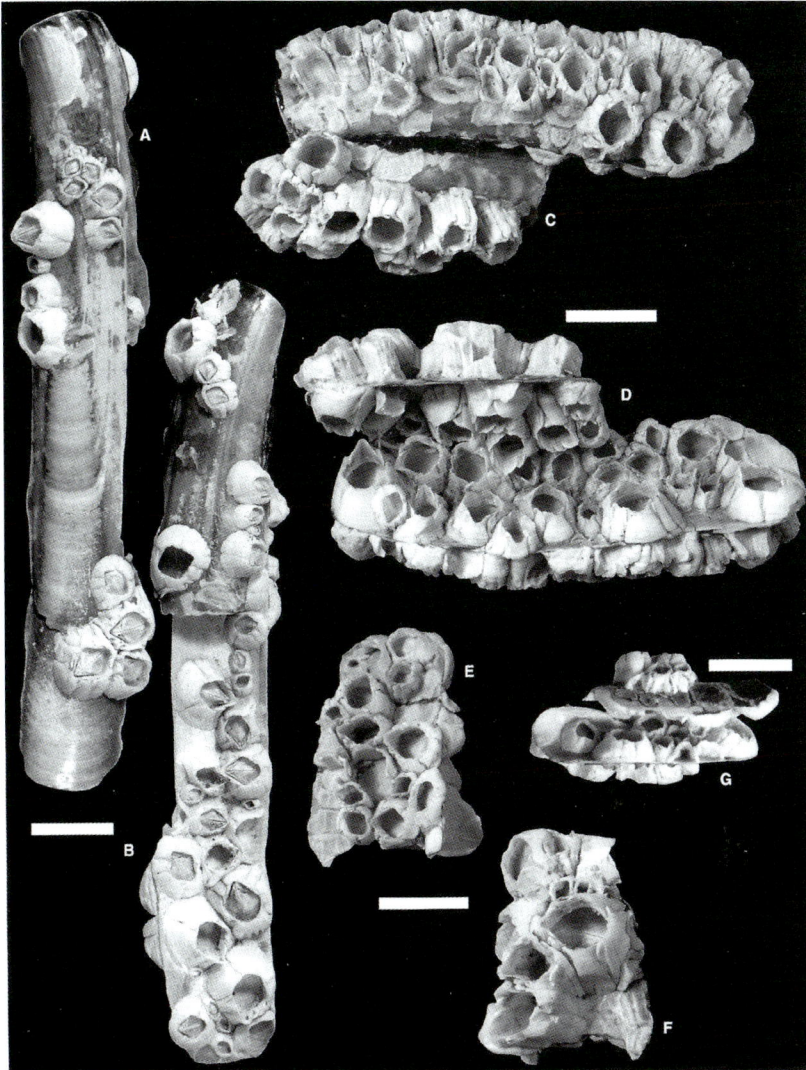

**Figure 11.3** Recent razor shells, *Ensis americanus* (Binney), mainly articulated (except **E**, **F**), and encrusted by balanid barnacles (*Balanus crenatus* Brugière), from Zandvoort, Noord-Holland, The Netherlands (after Donovan, 2007, text–fig. 2). Specimens are deposited in the Naturalis Biodiversity Center, Leiden, the Netherlands (prefix RGM). (**A**, **B**) RGM 211 470, near-complete right valve (**A**) articulated with about 40 per cent of left valve (**B**). Balanids close-packed internally, but forming small, broadly spaced clusters externally. (**C**, **D**) RGM 211 471, incomplete shell formed from two broken valves, still articulated and with densely packed balanids on the outer (**C**) and inner (**D**) surfaces of both valves. (**E**, **F**) RGM 211 473, fragment of one valve encrusted on inner and outer surfaces by balanids, that is, a balanulith *sensu* Cadée (2007); note that barnacles are commonly larger on the outer surface (**F**), unlike RGM 211 471 (**C**, **D**). (**G**) RGM 211 472, incomplete shell formed from two broken valves, still articulated and with densely packed balanids on the outer (upper and lower) and inner surfaces (centre) of both valves. Scale bars represent 10 mm.

**Figure 11.4** Broken oyster valve, *Ostrea edulis* Linné, RGM 791 569, encrusted by oysters and *Balanus* sp. cf. *B. perforatus* Bruguiére and bored by annelids (after Donovan & Novak, 2015, fig. 1). (**A, B**) Outer (**A**) and inner surfaces (**B**) of broken valve showing encrustation by oysters and balanids externally and balanids internally. The feature towards 10–11 o'clock from the left-hand balanids in (**B**) is the spionid polychaete boring *Caulostrepsis taeniola* Clarke. (**C**) Detail of balanids and *C. taeniola* on inner surface; note small round holes, *Oichnus simplex* Bromley, near the bases of some balanids. (**D**) Balanids at commissure on inner surface and encrusted, in turn, by an oyster. (**E**) Balanids encrusting inner surface; note common *O. simplex* low on the shells. Scale bars represent 10 mm. Specimens not coated for photography.

multiple specimens of juvenile oysters and balanid barnacles, *Balanus* sp. cf. *B. perforatus* Bruguière. At least nine barnacles bear drill holes, commonly singly, of *Oichnus simplex* Bromley, most probably produced by a muricid gastropod. This predator attacked the balanid shells close to the base of the cone, thus avoiding the attention of the cirri.

## Guides to identification

For further discussion and guides to identification of the extant British barnacles, see Rainbow (1984) and Southward (2008).

## References

Cadée, G.C. (2007) Balanuliths: Free-living clusters of the barnacle *Balanus crenatus*. *Palaios*, **22**: 680–681.

Collins, J.S.H., Donovan, S.K. & Mellish, C. (2014) An illustrated guide to the fossil barnacles (Cirripedia) from the Crags (Plio-Pleistocene) of East Anglia. *Proceedings of the Geologists' Association*, **125**: 215–226.

Donovan, S.K. (1988) Palaeoecology and taphonomy of barnacles from the Plio-Pleistocene Red Crag of East Anglia. *Proceedings of the Geologists' Association*, **99**: 279–289.

———. (1989a) Palaeoecology and significance of barnacles in the Pliocene *Balanus* bed in Tobago, West Indies. *Geological Journal*, **24**: 239–250.

———. (1989b) Taphonomic significance of the encrustation of the dead shell of Recent *Spirula spirula* (Linné) (Cephalopoda: Coleoidea) by *Lepas anatifera* Linné (Cirripedia: Thoracia). *Journal of Paleontology*, **63**: 698–702.

———. (1999) Pumice and pseudoplankton: Geological and paleontological implications of an example from the Caribbean. *Caribbean Journal of Science*, **35**: 323–324.

———. (2007) A cautionary tale: Razor shells, acorn barnacles and palaeoecology. *Palaeontology*, **50**: 1479–1484.

———. (2011) Post-mortem encrustation of the alien bivalve *Ensis americanus* (Binney) by the barnacle *Balanus crenatus* Brugière in the North Sea. *Palaios*, **26**: 665–668.

———. (2025) A beachcomber's field guide to The Palisadoes, Jamaica. *Bulletin of the Mizunami Fossil Museum*, **51**: 67–74.

Donovan, S.K., Cotton, L., Ende, C. van den, Scognamiglio, G. & Zittersteijn, M. (2014) Taphonomic significance of a dense infestation of *Ensis americanus* (Binney) by *Balanus crenatus* Brugière, North Sea. *Palaios*, **28** (for 2013): 837–838.

Donovan, S.K. & Novak, V. (2015) Site selectivity of predatory borings in Late Pliocene balanid barnacles from south-east Spain. *Lethaia*, **48**: 1–3.

Miller III, W.M., & Brown, N.A. (1979) The attachment scars of fossil balaenids. *Journal of Paleontology*, **53**: 208–210.

Pickerill, R.K. & Donovan, S.K. (1998) Ichnology of the Pliocene Bowden shell bed, southeast Jamaica. In Donovan, S.K. (ed.), *The Pliocene Bowden Shell Bed, Southeast Jamaica. Contributions to Tertiary and Quaternary Geology*, **35**: 161–175.

Rainbow, P.S. (1984) An introduction to the biology of British littoral barnacles. *Field Studies*, **6**: 1–51.

Southward, A.J. (2008) Barnacles. *Synopses of the British Fauna* (new series), **57**: viii+140 pp.

Wignall, P.B. & Simms, M.J. (1990) Pseudoplankton. *Palaeontology*, **33**: 359–378.

# CHAPTER 12

## Gastropods

### Form and function

The gastropods, or snails, are a class of univalve molluscs (= Class Gastropoda), which typically have a spirally coiled shell, although this is secondarily lost in some taxa, such as the terrestrial snails known as slugs. The gastropods are highly successful. At the present, gastropods are widespread and diverse in the marine, freshwater and terrestrial environments. They range back at least as far as the Early Cambrian.

The gastropod soft parts consist of a large foot (Fig. 12.1) which has a well-differentiated head at the anterior end that bears eyes and tentacles. The mouth (Fig. 12.1) bears a radula, a rasping organ comprised of many tooth-like elements. The elongate foot is an organ of locomotion, propelling the animal forwards by repeated waves of muscular contraction. These have the effect of moving the 'sole' of the foot across the substrate. The anus and gills open above the head. This is because the gastropod visceral mass has undergone torsion. Consider the geometry of a larval snail, in which the anus opens posteriorly into the mantle cavity, which was also the site of the gills. During development, the position of the mantle cavity is twisted through 180° (= torsion). Why torsion occurs is problematic. It is not related to the coiling of the shell. The most probable explanation suggests that relocation of the mantle cavity to an anterior position enables retraction of the soft parts into the protection of the shell by providing space for this movement.

Respiration in aquatic gastropods is by the gills which occur in the mantle cavity (Fig. 12.1). In some marine taxa a tubular extension of the mantle, called the inhalant siphon, acts to direct a flow of water over the gills. In species where an inhalant siphon is present, the anterior of the shell is modified into a more or less extended siphonal canal (Fig. 12.2). In some species there is a narrow slit in the exterior lip of the aperture which is called the exhalant slit or slit-band; this marks the position where the faeces are passed out. As the shell grows, this slit is progressively infilled by a banded structure called the selenizone (Fig. 12.3).

The larval shell is bilaterally symmetrical, forming a planispiral (= flat spiral) coil. Some ancient molluscs of uncertain classification (either a gastropod or a member of a minor group, the monoplcophorans), such as *Bellerophon* (see, for example, Owen & Harper, 1996, pl. 24, figs 3, 4), retained a planispiral shell in adulthood. All modern gastropods have an asymmetrical shell (apart from some groups in which bilateral symmetry has been

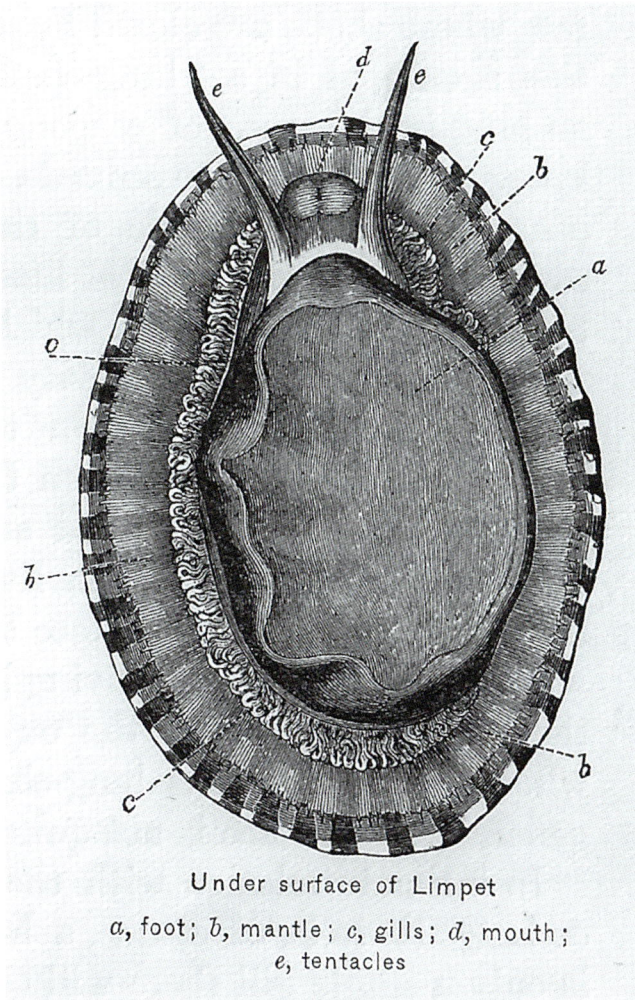

**Figure 12.1** The foot and other soft tissues of a limpet (after Step, 1901, p. 45). Limpets are gastropods with a conical shell (= patelliform) and a particularly strong, muscular foot, adapted to inhabit rocky substrates in high energy situations such as intertidally.

Under surface of Limpet

*a*, foot; *b*, mantle; *c*, gills; *d*, mouth; *e*, tentacles

secondarily reacquired or the shell is lost). However, a planispiral shell is bulky, as each successive whorl must enclose the previous whorl. In gastropods that have a spirally coiled, conical growth form, the shell twists about a central axis with each successive coil, or whorl. Whorls are thus built up successively, one on top of another (Fig. 12.2). Such shells are compact and may approach a globular outline.

The anterior opening of the shell is called the aperture (Fig. 12.2; also note the range of apertural forms shown in Clarkson, 1998, fig. 8.18) and is the point of emergence of the soft tissues. The soft tissues can be retracted into the shell. Many species have a lid-like structure, the operculum (Fig. 12.4), attached to the foot, which acts to seal off the aperture once the soft tissues have been withdrawn. The aperture bears an inner and an outer lip, which are often highly modified in adult gastropods. Lip modification is commonly indicative of the cessation of growth. The shell can be decorated in a number of ways, particularly by growth lines parallel to the outer lip of the aperture or ribs spiralling down the shell (Figs 12.2, 12.3).

**Figure 12.2** Apertural view of the whelk *Buccinum undatum* (Linné). The shell has a thickened inner and thin outer lips. The siphonal canal is at the bottom of the shell. Note the faint growth lines, parallel to the outer lip, and ribbing. Specimens collected from the beach at Southport, Merseyside, north-west England. Scale in cm and mm.

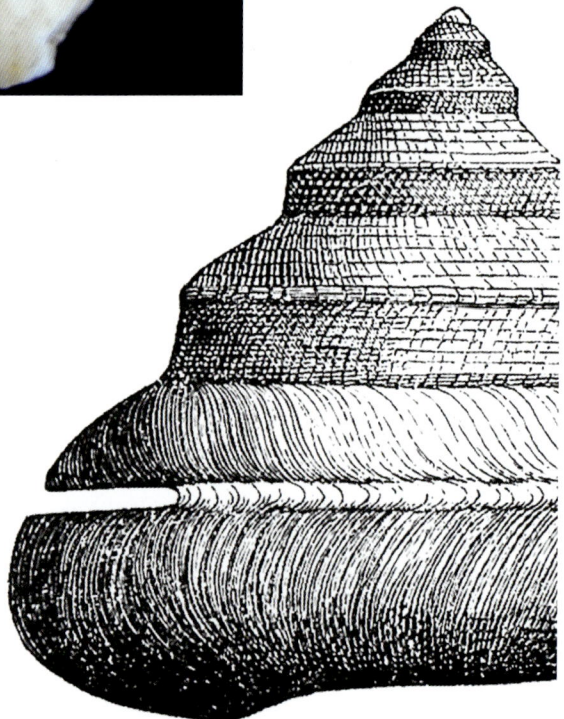

**Figure 12.3** Incomplete view of the Jurassic gastropod *Pleurotomaria subscalaris* Deslongchamps viewed from behind the aperture (after von Zittel, 1927, fig. 852B). The exhalent slit-band is to the left; this is confluent with the selenizone, which trends right.

**Figure 12.4** Opercula of Late Pleistocene land snails from the Red Hills Road Cave, parish of St. Andrew, Jamaica (after Paul & Donovan, 2006, pl. 17); those of marine gastropods are broadly similar. (**1, 2**) *Fadyenia jayana* (C.B. Adams), external surfaces of isolated opercula. (**1**) 0.7 mm maximum dimension. (**2**) 1.0 mm maximum dimension. (**3, 4**) *Fadyenia lindsleyana* (C.B. Adams). (**3**) Oblique umbilical view of shell with operculum in place, 2.3 mm maximum dimension. (**4**) Detail of operculum of preceding example, 1.0 mm maximum dimension. All scanning electron micrographs of specimens coated with 60 per cent gold-palladium.

In most shells there is a central structure, the columella (Fig. 12.5), which marks the position of the coiling axis. In some species the shell is broadly triangular so that a columella is not developed. Instead, the inner walls of the shell define a conical cavity.

The spire (Figs 12.2, 12.5) is the region of the shell above the final whorl which represents the youngest part of the gastropod valve; the spire terminates in the pointed apex. The shell is composed of calcium carbonate, commonly aragonite. In most gastropods the shell coiling direction is dextral, that is coiling to the right (Fig. 12.2). However, in certain species, left-handed (= sinistral) coiling is the norm (see, for example, Todd & Parfitt, 2017, pl. 40, fig. 7).

Many marine gastropods are herbivorous, grazing on algae. However, some species have adapted to a carnivorous habit, employing a variety of adaptations to subdue and devour

**Figure 12.5** Broken shells of *Buccinum undatum* (Linné) showing the central columella and spire. Specimens collected from the beach at Southport, Merseyside, north-west England. Scale in cm and mm.

prey. One method of feeding that has left a good fossil record involves rasping a circular, conical boring through the shell of a prey organism to access the edible tissues (Fig. 12.6). This form of predation is typical of the naticid gastropods.

## On the beach

Fossil molluscs are a common feature of shell accumulations of beaches everywhere, but groups tend to be preserved in a loosely hierarchical order. In my experience, bivalves are the commonest molluscs on the great majority of beaches (see, for example, Chapter 17; Donovan, 2025). They have two valves (that is, after death they produce two disarticulated valves rather than just one as does a gastropod) and are herbivorous, most commonly living infaunally in their hundreds. The univalved gastropods are rarer, commonly representing predators that preyed on the bivalves; in any ecosystem, prey (in this case, bivalves) must outnumber predators (Ager, 1963, fig. 1.2). Scaphopods (also univalves) and chitons are less diverse than either bivalves or gastropods. Chitons are most commonly preserved as disarticulated valves except in rare sites. Shelled cephalopods may float after death and can be concentrated on a beach due to a major storm event (Jongbloed *et al.*, 2016).

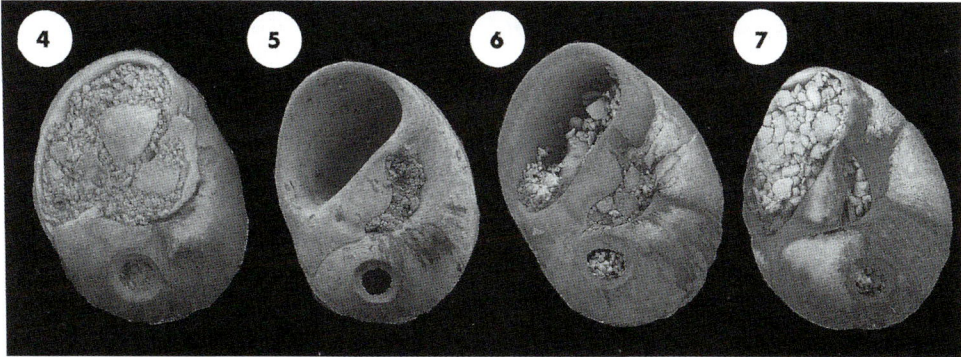

**Figure 12.6** Four specimens of the Pliocene gastropod *Natica castrenoides* Woodring (after Pickerill & Donovan, 1998, pl. 3, figs 4–7). Note the site specificity of the small round boring, *Oichnus paraboloides* Bromley. Maximum dimensions 5.2 mm (**4**); 6.4 mm (**5**); 5.3 mm (**6**); and 5.6 mm (**7**). Bowden shell beds, Bowden Member, south-east Jamaica. Scanning electron micrographs of specimens in the collection of the University of New Brunswick, Fredericton, Canada. All specimens coated with 60 per cent gold-palladium.

It is normal to collect well-preserved shells, but broken shells can reveal features such as the columella (Fig. 12.5). Donovan (2021) discussed a shell of *Buccinum undatum* with a broken outer lip to the aperture. This may have been due to natural breakage, but it could equally be the result of predation by a crab, peeling the shell back with its claw.

All dead shells on the beach are a worthy target for encrusting organisms, such as balanid barnacles, tube-dwelling worms, bryozoans and encrusting bivalves, particularly oysters (Donovan, 2013). Take care to note where encrustation occurs. If barnacles are concentrated externally and the inside of the shell is clean, then it may represent an infestation while the gastropod was alive. Any encrusting organisms within the shell must have arrived after the death of the snail.

Borings invite analysis (Donovan, 2021), particularly those that provide evidence of predation or even cannibalism. In Figure 12.6, four naticid snails show borings of similar form and situation. These penetrative borings are *Oichnus paraboloides* Bromley and are typical of borings made by naticid snails. Naticids are known to be cannibals. So, in this fossil example, the naticids may have been feeding on other naticids.

## Guides to identification

General guides to the seashore life of the British Isles are many; a visit to your local bookshop will probably yield one, probably more. My favourite is Barrett & Yonge (1977), but this is getting a little long in the tooth. Concerning gastropods and other molluscs, some more specialised texts of similar vintage, but still available second-hand (see abebooks.com), include McMillan (1977) and Beedham (1972). When I'm on the west coast of the Atlantic Ocean and in the Caribbean I use Morris (1975) and Warmke & Abbott (1961). But wherever you are, there will be inexpensive books on the local marine Mollusca.

# References

Ager, D.V. (1963) *Principles of Paleoecology.* McGraw-Hill, New York.

Barrett, J. & Yonge, C.M. (1977) (first published 1958) *Collins Pocket Guide to the Sea Shore.* Collins, London.

Beedham, G.E. (1972) *Identification of the British Mollusca.* Hulton Educational Publications, Amersham.

Clarkson, E.N.K. (1998) *Invertebrate Palaeontology and Evolution.* 4th ed. Blackwell Science, Oxford.

Donovan, S.K. (2013) An unusual association of a Recent oyster and a slipper limpet. *Deposits,* **35**: 5.

———. (2021) Fossils explained 81: An exciting, but bored whelk. *Geology Today,* **37**: 194–197.

———. (2025) A beachcomber's field guide to the other side of Doggerland: Zandvoort aan Zee, Noord Holland, The Netherlands. *Bulletin of the Geological Society of Norfolk,* **75**: 45–55.

Jongbloed, C.A., Gier, W. de, Ruiten, D.M. van & Donovan, S.K. (2016) Aktuo-paläontologie of the common cuttlefish, *Sepia officinalis,* an endocochleate cephalopod (Mollusca) in the North Sea. *PalZ,* **90**: 307–313.

McMillan, N.F. (1977) *The Observer's Book of Seashells of the British Isles.* Frederick Warne, London.

Morris, P.A. (Clench, W.J., ed.) (1975) *A Field Guide to Shells: Atlantic and Gulf Coasts and the West Indies.* Houghton Mifflin, Boston.

Owen, A.W. & Harper, D.A.T. (1996) Other molluscs. In Harper, D.A.T. & Owen, A.W. (eds), *Fossils of the Upper Ordovician*: 130–137. Palaeontological Association Field Guides to Fossils, 7. London.

Paul, C.R.C. & Donovan, S.K. (2006) Quaternary land snails (Mollusca: Gastropoda) from the Red Hills Road Cave, Jamaica. *Bulletin of the Mizunami Fossil Museum,* **32** (for 2005): 109–144.

Pickerill, R.K. & Donovan, S.K. (1998) Ichnology of the Pliocene Bowden shell bed, southeast Jamaica. In Donovan, S.K. (ed.), *The Pliocene Bowden Shell Bed, southeast Jamaica: Contributions to Tertiary and Quaternary Geology,* **35**: 161–175.

Step, E. (1901) *Shell Life: An Introduction to the British Mollusca.* Frederick Warne, London.

Todd, J. & Parfitt, S. (eds) (2017) *British Cenozoic Fossils (Paleogene, Neogene and Quaternary).* 6th ed. Natural History Museum, London.

Warmke, G.L. & Abbott, R.T. (1961) *Caribbean Seashells.* Livingston Publishing, Narberth, PA.

Zittel, K.A. von (Eastman, C.R., ed.). (1927) *Text-book of Paleontology.* vol. 1. 2nd ed. Macmillan, London.

# CHAPTER 13

# Bivalves

## Form and function

The bivalve molluscs (otherwise called lamellibranches, pelecypods or clams) are a group in which the shell is comprised of an articulated pair of valves which enclose the internal soft tissues. The infaunal bivalves are usually equivalve and bilaterally asymmetrical, facilitating burrowing; epifaunal bivalves such as oysters, scallops and mussels are commonly inequivalve (Fig. 13.1).

The bivalve shell is composed of calcium carbonate, either aragonite or calcite or both. The valves articulate at the hinge in the region of the umbo. Articulation is by a system of interlocking teeth and sockets. These show a great variety of form and are of importance in classification (see Skelton, 1985, table 6.4.1).

The umbo is not usually central, but is displaced towards one end, which is anterior by convention (Fig. 13.2). The umbonal surface is considered dorsal. The posterior and ventral regions, respectively, are opposite to these. To determine the left from the right valve, place an articulated shell in your cupped hands, with umbo uppermost and pointing away from you. The left valve is now in your left hand and the right valve is in your right hand.

Bivalves open their valves with a spring-like ligament, situated in the umbonal region. The ligament can be either external or internal. If external, it occurs between the hinge and the umbo in a region called the cardinal area. The valves are closed by adductor muscles. These are paired in most bivalves, but are single in, for example, oysters and scallops. Muscle scars are prominent features on the insides of valves (Fig. 13.2).

External growth lines are concentric around the umbo. Radiating ribs usually originate from the umbo. Some epifaunal bivalves, particularly oysters, are spiny.

The interior of the shell is covered by the mantle, which secretes the valves. The join between the mantle lobes and the valves is called the pallial line (Fig. 13.2). In epifaunal and shallow infaunal bivalves the pallial line is parallel to the commissural margin of the shell. However, in deeply infaunal species, the pallial line shows a fold called the pallial sinus. This opens into the posterior half of the shell and is only found in those species with a retractable paired siphon, marking the position to which it can be withdrawn. The paired siphon of infaunal bivalves is an extension of the mantle which keeps the animal in contact with the surface. One siphon is incurrent, through which water is drawn into the shell for

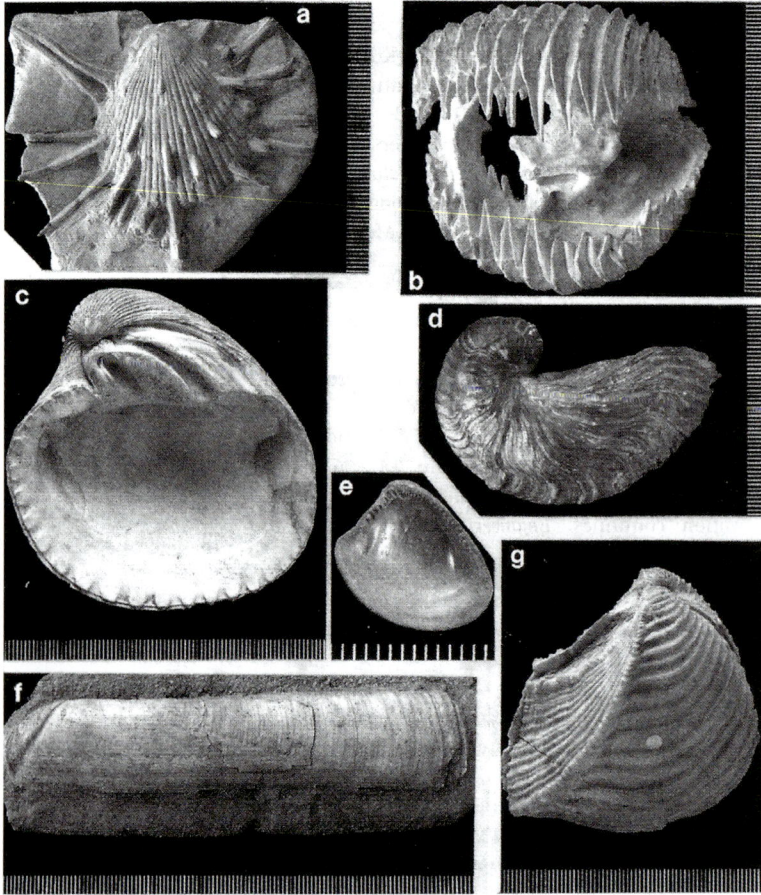

**Figure 13.1** (After Hensley & Donovan, 2005, fig. 1.) Something of the diversity of bivalves; specimens in the Natural History Museum, London. (**a**) The spiny oyster *Spondylus spinosa* (J. Sowerby), Upper Cretaceous, Chalk, England. The prominent spines would have provided stability on the substrate; thus, this is the lower valve. (**b**) *Arctostrea carinatum* (Lamarck), Cenomanian, Lower Grey Chalk, Dover, Kent. The interlocking zigzag commissure provided both protection against predation and delimited water currents flowing over the gills. (**c**) *Venericor planicostata* (Lamarck), Eocene, Bracklesham, Sussex. Inner surface of right valve. The huge cardinal teeth and adductor muscle scars joined by the pallial line are clearly visible. (**d**) *Gryphaea arcuata* Lamarck, Lower Jurassic, Fretherne, Gloucestershire. These inequivalve Jurassic oysters are popularly known as 'Devil's toenails'. They may be preserved in life position with the larger valve, seen here from the side, embedded in the substrate, and the smaller valve, concealed in this view, free of the sediment and opening like a lid. The growth lines are prominent. (**e**) *Nucula similis* var. *trigona* J. Sowerby, Upper Eocene, Barton, Hampshire. Inside of the right valve showing taxodont dentition. The inside of the shell is nacreous. (**f**) The razor shell *Solena plagiaulax* (Cossman), Middle Eocene, Bracklesham Bay, Sussex. The smooth, elongate valves would have facilitated burrowing vertically into sediment. (**g**) *Trigonia reticulata* Agassiz, Lower Kimmeridge Clay, near Weymouth, Dorset. Concentric growth lines and radial ribs are prominent in different parts of the valve. All gradations in mm.

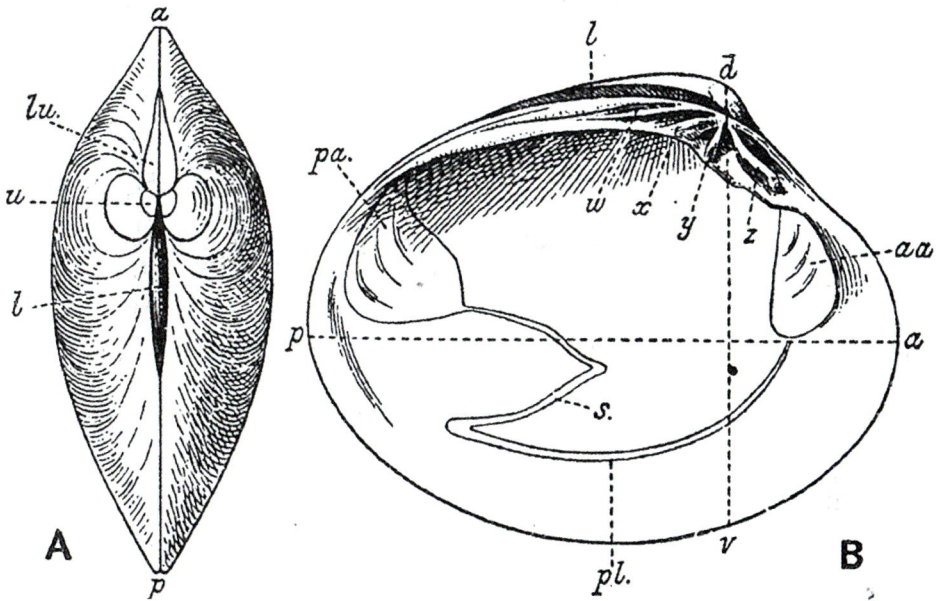

**Figure 13.2** The Recent bivalve *Callista chione* (Linné) (after Woods, 1955, fig. 123). (**A**) Dorsal view of the two articulated valves. (**B**) Interior of left valve, maximum dimension 90 mm. Key: *a* = anterior border; *p* = posterior; *d* = dorsal; *v* = ventral; *lu* = lunule; *u* = umbo; *l* = ligament; *aa* = anterior adductor muscle scar; *pa* = posterior adductor muscle scar; *pl* = pallial line; *s* = pallial sinus; *w, x, y* = cardinal teeth; *z* = anterior lateral tooth.

feeding and respiration. Water containing metabolic waste is carried back to the surface through the excurrent siphon.

The foot and gills of bivalves both extend outside the central visceral palps. The foot is a muscular organ used as a digging tool by infaunal bivalves. The comb-like gills draw water into the shell by creating currents driven by cilia. The gills are respiratory and also trap food particles in mucus, which are passed to the mouth by the cilia.

## On the beach

Bivalves are aquatic and, like gastropods, are common in both marine and freshwater settings; unlike gastropods, they are not terrestrial. In many environments, bivalves are common and diverse. The one group of invertebrates that I always expect to find on any beach is the bivalves. Recently dead shells will be articulated by the ligament, which breaks eventually, but particularly when subaerially exposed so that it can dry and become brittle (Figs 2.4, 13.2). About 75 per cent of marine bivalve taxa are infaunal, both burrowers and borers; they are distinct in form from most epifaunal species.

*Infaunal burrowers* (Figs 13.1c, e-g, 13.2–13.4): Infaunal bivalves are free-living and not attached by a byssus or by cementation. Most are soft-sediment burrowers, pulling themselves down into the substrate with their muscular foot and feeding from the surface

**Figure 13.3** Some common seashells (after Step, 1901, plate opposite p. 142). All right valves, all burrowers except (**4**). (**1**) Smooth cockle, *Cardium crissum* (Gmelin). Maximum dimension 64 mm. (**2**) Large sunset-shell, *Gari depressa* (Pennant). Maximum dimension 67 mm. (**3**) Ribbed sunset-shell, *Gari tellinella* (Lamarck). Maximum dimension 19 mm. (**4**) The boring oval piddock, *Zirfaea crispata* (Linné). Maximum dimension 52 mm.

waters using the paired siphons. Shallow burrowers lack the pallial sinus that is present in deep burrowers (Fig. 13.2), the latter having large siphons that fill the extra space if retracted. Very deep burrowers also have a permanent gape; that is, even when the valves are 'closed', there is still a broad opening through which the siphons protrude.

As one example of a common group of shallow-water, burrowing bivalves, razor shells demonstrate how to make a contribution to your knowledge of a particular group of bivalves or, indeed, any group of seashore invertebrates. Razor shells can be present in great numbers on some beaches, presumably excavated from the seafloor sediment and carried onshore during storms. My interest in them was fired by a chance observation of one specimen (at first) that showed an unusual pattern of preservation. Spotting rarities amongst the common requires careful observation and an awareness of what questions might be asked if something unusual is observed. Alternatively, some taxa are rare, which asks an obvious question 'why'? Watching the same beach through an annual cycle, year after year, will make you sensitive to changes other than those that repeat annually. Because bivalves are so common, they are likely to be particularly sensitive to change and pose interesting questions, providing you are likewise sensitive to them. At the same time, remember you are a palaeontologist and the sort of questions that you ask will not be those of the zoologist (e.g., Donovan *et al.*, 2020, and references therein).

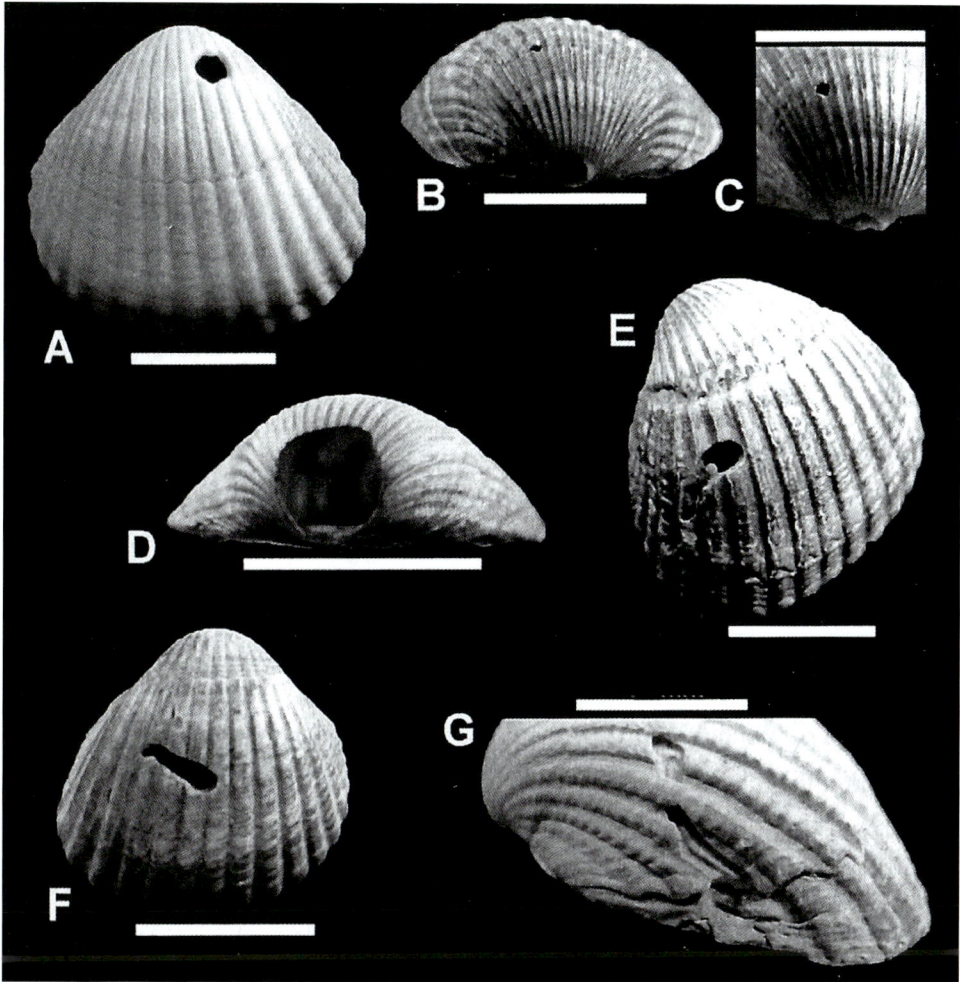

**Figure 13.4** (After Hoogduin *et al.*, 2017, fig. 3.) Valves of the common edible cockle *Cerastoderma edule* (Linné) collected from the beach at Zandvoort aan Zee, North Sea coast, the Netherlands, showing various styles of preservation. Specimens are deposited in the Naturalis Biodiversity Center, Leiden, the Netherlands (prefix RGM), unless stated otherwise. (**A**) Hexagonal-shaped *Oichnus paraboloides* Bromley, a predatory boring in a left valve; specimen lost. (**B, C**) RGM 792 296, left valve with a small, predatory *O. simplex* boring close to the broken umbo. (**D**) RGM 792 298, a large, rounded hole on a mechanically damaged umbo of a right valve. (**E**) RGM 792 301, right valve bored by polychaete annelid worms, *Caulostrepsis taeniola* Clarke. (**F**) RGM 792 297, perforation in right valve caused by collapse of a *Caulostrepsis*. (**G**) RGM 792 302, right valve with particularly well-preserved *C. taeniola*. Specimens uncoated; scale bars represent 10 mm.

As an example of potential studies, remember that bivalves are the prey organisms for many invertebrate (annelids, gastropods, crabs) and vertebrate (fishes, birds, man) organisms. A mass shell deposition event is likely to be due to storms or other environmental perturbations, yet in amongst this bivalve bonanza there is likely to be evidence for cause of death in at least some specimens, such as borings by predatory gastropods (Pickerill & Donovan, 1998; Fig. 13.4A–C herein). Similarly, an unusual pattern of encrustation, boring or breakage may be notable as a unique specimen or as part of a large collection from the same site and time (Fig. 13.4).

***Infaunal borers*** (Figs 13.3.4, 13.5): Some bivalves are hard substrate borers in wood or rock (limestones and mudrock). These may or may not gape broadly and bore partly by chemical secretions and partly by rocking their thickened calcareous valves against the substrate. Boring is a one-way process. Boring only occurs at the anterior end of the shell. As the bivalve bores deeper, it grows larger, the deeper borehole is necessarily broader and

**Figure 13.5** The common piddock, *Pholas dactylus* Linné, in its club-shaped boring (after Step, 1901, fig. on p. 164; see also Tebble, 1976, pp. 179–180, text–fig. 94). The shell is 67 mm in maximum dimension.

club-shaped, and the mollusc is unable to retreat down its narrower, juvenile conduit. The boring thus produced is the distinctive *Gastrochaenolites* Leymerie (Fig. 13.5).

Collecting shells on the beach, there is no easy method to distinguish burrowers from borers and it is easiest to just memorise the few species of the latter that you are likely to find (Tebble, 1976). On most beaches, burrowers are common, and borers are rare. Borers may have a wide gape, but so, too, would have a deep burrower, such as a razor shell. It is only on rare beaches that borers might be the dominant infaunal bivalve specimens (e.g., Chapter 17).

Sometimes, if you are lucky, a bored clast on the beach preserves the borer *in situ*. A genuine borer in a boring should be easy to determine. Other invertebrates, most commonly nestling bivalves, bryozoans and tube-forming worms, may occupy a boring after the death of the borer. When live and articulated, a boring bivalve cannot get out of its borehole; after death and disarticulation of the shell, it is lost more easily, particularly from a clast being rolled on the sea floor.

***Epifaunal bivalves*** (Figs 2.4, 13.1a, b, d, 13.6, 13.7): Three groups are considered herein to demonstrate the range of morphologies shown by epifaunal bivalves. Mussels (Fig. 13.6) are a group of kidney-shaped bivalves which attach to the substrate by a bundle of threads called a byssus, which is produced by the foot. The position of the byssus in fossil mussels can be determined by recognising the gap in the commissure through which it emerged,

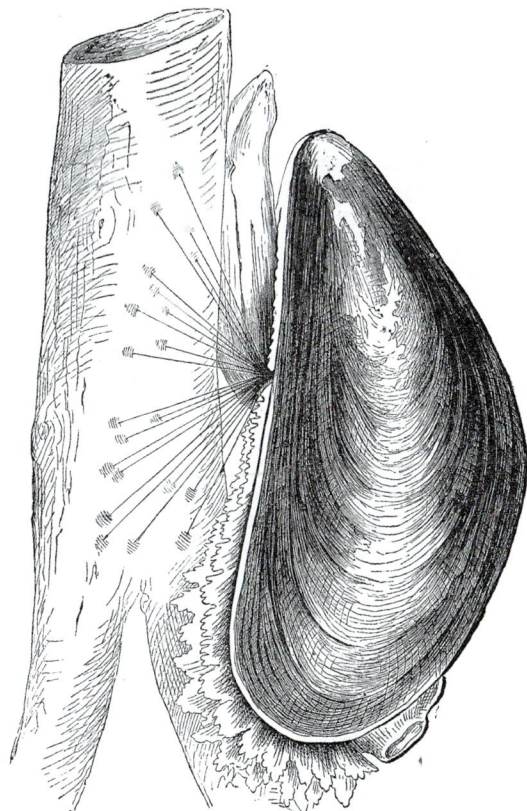

**Figure 13.6** The common mussel, *Mytilus edulis* Linné, feeding and attached by the byssus (after Step, 1901, fig. on p. 66; see also Tebble, 1976, pp. 40, 41, 43, pl. 3, figs a, b). The shell is 87 mm maximum dimension.

**Figure 13.7** Large, but incomplete, recumbent shell of the oyster *Crassostrea virginica* (Gmelin) *in situ* in the Neogene oyster bed at Farquhar's Beach, Jamaica (after Littlewood & Donovan, 1988, pl. 91, fig. 8). Specimen about 367 mm long. This species is extant.

the byssal notch. A mussel is seemingly like an infaunal bivalve to the casual observer, but the commissure in the region of the byssus is flexed, making the shell inequivalve.

Scallops (or pectinaceans) are inequivalve and almost bilaterally symmetrical. The regions around the umbo are extended into a pair of 'ears' and a byssus arises in this region. In many species, the byssus is lost in maturity, and the bivalve becomes free-living and capable of swimming. Normal swimming entails clapping both valves together and directing water out of the 'ears', so that the direction of movement is ventrally. An escape reaction involves violent clapping of the valves, forcing water out of the commissure, so that movement is in a dorsal direction.

Oysters are asymmetrical and inequivalve (Figs 2.4, 13.1a, b, d, 13.7). Oysters follow two growth strategies. Many of them cement the lower valve to a hard substrate, so gaping is by raising the unattached (free) valve. Other oysters are 'mud stickers', living free on the sediment surface, but being stable due to their large size and weight. Some species are further stabilised by spines (Fig. 13.1e). Certain oysters are amongst the largest living and fossil bivalves (e.g., Fig. 13.7).

## Guides to identification

The key book that you will need to identify bivalves around the British Isles is Tebble (1976). A little long in the tooth, perhaps, but full of meat and with first-rate diagrams. The excellent Beedham (1972) extends the coverage to freshwater bivalves, if they are also of interest.

McMillan (1977) is good value for money and fits even tiny coat pockets. Good books on the seashore with good sections on molluscs are many and include Barrett & Yonge (1977) and Campbell (1976), amongst others. And for the fun of it read Street (2019).

## References

Barrett, J. & Yonge, C.M. (1977) (first published 1958) *Collins Pocket Guide to the Sea Shore*. Collins, London.

Beedham, G.E. (1972) *Identification of the British Mollusca*. Hulton Educational Publications, Amersham.

Campbell, A.C. (1976) *The Country Life Guide to the Seashore and Shallow Seas of Britain and Europe*. Country Life Books, London.

Donovan, S.K., Hoeksema, B.W., Fransen, C.H.J.M., Vonk, R. & Adema, J.P.H.M. (2020) Unusual preservation of North Sea shells: Scheveningen, North Sea coast, the Netherlands. *Bulletin of the Geological Society of Norfolk*, **70**: 55–65.

Hensley, C. & Donovan, S.K. (2005) Fossils explained 49: bivalves. *Geology Today*, **21**: 71–75.

Hoogduin, A.L., Visscher, M.R. & Donovan, S.K. (2017) Aspects of the neotaphonomy of three species of bivalve molluscs common in the North Sea. *Bulletin of the Geological Society of Norfolk*, **66**: 3–17.

Littlewood, D.T.J. & Donovan, S.K. (1988) Variation of Recent and fossil *Crassostrea* in Jamaica. *Palaeontology*, **31**: 1013–1028.

McMillan, N.F. (1977) *The Observer's Book of Seashells of the British Isles*. Frederick Warne, London.

Pickerill, R.K. & Donovan, S.K. (1998) Ichnology of the Pliocene Bowden shell bed, southeast Jamaica. In Donovan, S.K. (ed.), *The Pliocene Bowden Shell Bed, Southeast Jamaica. Contributions to Tertiary and Quaternary Geology*, **35**: 161–175.

Skelton, P.W. (1985) Bivalvia. In Murray, J.W. (ed.), *Atlas of Invertebrate Macrofossils*. Longman (for the Palaeontological Association), Harlow, 81–100.

Step, E. (1901) *Shell Life: An Introduction to the British Mollusca*. Frederick Warne, London.

Street, P. (2019) *Shell Life on the Seashore*. New edition. Faber & Faber, London.

Tebble, N. (1976) *British Bivalve Seashells: A Handbook for Identification*. 2nd ed. HMSO, Edinburgh.

Woods, H. (1955) *Palaeontology: Invertebrate*. 8th ed. Cambridge University Press, Cambridge.

# CHAPTER 14

## Cephalopods, scaphopods and chitons

### Form and function

In this chapter I look at those groups of molluscs that are encountered less often on beaches, namely the cephalopods (octopus, squid, etc.), scaphopods (elephant tusk shells) and chitons (mail shells). All three have their fascination and can pop up when least expected (e.g., Donovan *et al.*, 2001).

*Cephalopods* (Figs 14.1–14.3): Cephalopods are a group of free-swimming, univalved molluscs with the foot developed as a series of tentacles. A pair of highly developed eyes are present which show many similarities to those of vertebrates. Locomotion is by a system of

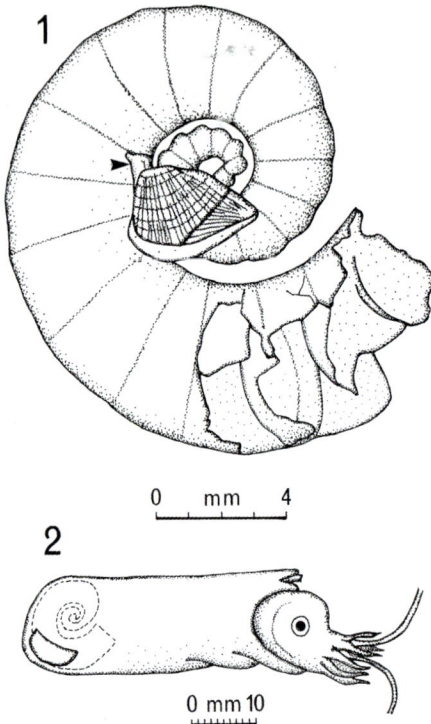

**Figure 14.1** *Spirula spirula* (Linné) (after Donovan, 1989, fig. 2). (**1**) The Natural History Museum, London, Zoology Department specimen 1988•373. Camera lucida drawing of a specimen encrusted by the goose barnacle *Lepas anatifera* Linné. The position of the barnacle's peduncle is indicated by an arrow. (**2**) Position of the shell in the living organism; the life orientation of *Spirula* is with the buoyant shell towards the surface. See also Figure 11.2.

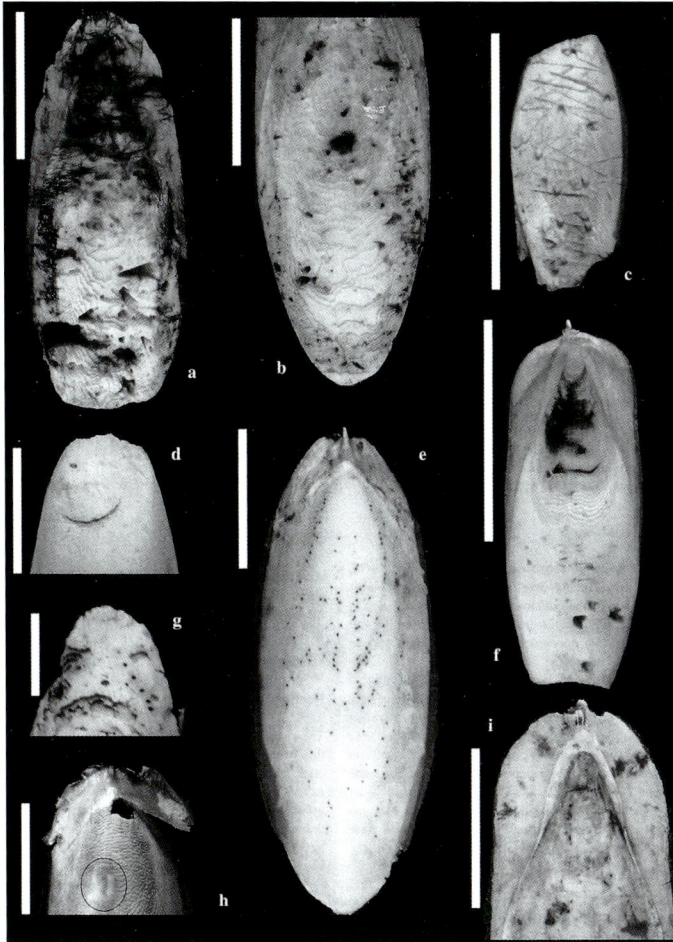

**Figure 14.2** Examples of Recent *Sepia officinalis* Linné cuttlebones collected from the beach at Katwijk, the Netherlands (Jongbloed *et al.*, 2016, fig. 3). The collective details of these shells shows that they represent a death assemblage of shells, washed ashore by a storm. (**a**) Naturalis Biodiversity Center, Leiden, the Netherlands RMNH.MOL.338218 showing several deep triangular pits (= seabird beak marks) and a dense accumulation of hair-like algae on the ventral surface. (**b**, **i**) RMNH.MOL.338219. (**b**) Showing a deep, rounded hole in the centre of the ventral surface (a damaged triangular pit?), associated with small triangular pits. (**i**) Showing some green 'stains' of algae apparent as dark patches on the ventral surface. (**c**) RMNH. MOL.338222 showing several scratches on the ventral surface (commonly occurring in pairs, forming V-shapes). (**d**) RMNH.MOL.338215 showing several marks on the ventral surface, most obviously the crescent-shaped bite-mark. (**e**) RMNH.MOL.338217, dorsal surface encrusted by numerous immature acorn barnacles *Balanus* sp. cf. *B. improvisus* (Darwin), apparent as dark dots on either side of the mid-line. (**f**) RMNH.MOL.338221 showing a large black stain towards the top of the ventral surface, above a black/brown line. (**g**) RMNH.MOL.338220 showing six little holes, sloping towards the lower right, oriented in a row on the (posterior) ventral surface. Scale bar represents 10 mm. (**h**) RMNH.MOL.338216 showing a raised 'lump' on the dorsal surface, inside the dark circle. Specimens uncoated. All scale bars represent 50 mm except (**g**).

jet-propulsion, water being expelled out of a funnel called the hyponome. Our knowledge of fossil cephalopods is based mainly upon shelled forms which contained a number of flotation chambers filled with gas and a single body cavity. The sole survivor of this form of cephalopod is the *Nautilus*. However, some modern forms retain an internal shell, such as *Spirula* (coiled; Fig. 14.1) and the cuttlefish *Sepia* (flattened; Fig. 14.2). The octopus and squid have completely lost all remnants of a shell.

Three major groups of cephalopods are recognised in the fossil record, the nautiloids (Fig. 14.3), the ammonoids and the coleoids. Ammonoids are extinct, but their fossils may be collected reworked on beaches in many parts of the country, such as Whitby and Folkestone. See an authoritative textbook on palaeontology for discussion of their palaeo-biology (such as Clarkson, 1998). Nautiloids are extant but not known from European waters. However, they have a good fossil record, and some readers may be lucky enough to find their shells on the beach in south-east Asia. Shelled coleoids, such as *Sepia* (Fig. 14.2), are the cephalopods most likely to be encountered on the British beach; again, refer to Clarkson (1998) for discussion of fossil coleoids, most particularly the Mesozoic belemnites, which, again, occur as reworked fossils on some beaches.

Nautiloids have a range from the Upper Cambrian to Recent. The shell is external, the animal inhabiting the body cavity at the broad (youngest) end. The opening of the body cavity is called the aperture. As the animal grew it sealed off successive body cavities by a

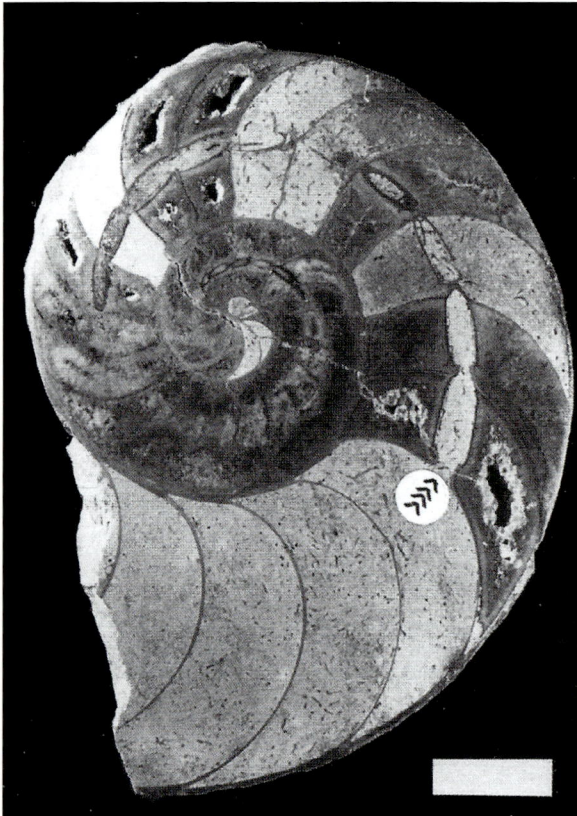

**Figure 14.3** Indeterminate fossil nautiloid from the Mesozoic of France (after Donovan, 2000, fig. 1A). Varnished section cut through a planispiral conch close to the plane of coiling. Separate chambers are infilled variously with sedimentary rock (pale tone) and sparry cement (dark tone). The cluster of arrows indicates the position of the central siphuncle (not cut in every chamber). Also note the concave-forward orientation of the septa and the absence of the body chamber (which would have been in the lower left), which has not been preserved. The septal necks support the siphuncle and are orientated away from the aperture; they are best seen adjacent to the arrows. Scale bar represents 20 mm.

series of walls, called septa (singular, septum). Old body cavities are thus modified to form enclosed chambers. In nautiloids the septa always have a concave surface towards the body cavity (Fig. 14.3). The gas-filled chambers act to give the animal buoyancy. Chambers are connected to each other and the body cavity by a tube called the siphuncle. The siphuncle is a conduit which is used to add or remove fluid to the chambers, thus altering the buoyancy and allowing the animal to move up or down in the water column. In each septum there is a small tube directed away from the body cavity and which forms part of the siphuncle, called the septal neck. The contact between the circumference of a septum and the inside of a shell is called the suture line. In nautiloids, the suture is always simple. The nautiloid shell is aragonitic and is therefore always replaced or lost (internal moulds) in fossil specimens. The nautiloid shell is primitively straight, becoming coiled in extant forms.

Differentiating between fossil nautiloids and ammonoids may be problematic. The following notes should aid in their differentiation.

Nautiloids

(1) Septa concave forwards.

(2) Siphuncle central.

(3) Septal necks point away from the aperture.

(4) Simple suture.

(5) Weak sculpture.

Ammonoids

(1) Septa convex forwards.

(2) Siphuncle ventral.

(3) Septal necks point towards the aperture.

(4) Complex suture.

(5) Strong sculpture.

The coleoids include all extant cephalopods, except for *Nautilus*, such as squids, octopuses, cuttlefishes and argonauts. The coleoid shell is internal, except where it has been lost completely. Although rare fossils of soft-bodied coleoids are known, their ancient record is based mainly upon the elongate, cigar-like hard parts of Mesozoic squid-like organisms called belemnites (Clarkson, 1998). In northern Europe, scour the beach for *Sepia* (Figs 14.2, 19.8), the so-called cuttlefish bones presented to parrots to sharpen their teeth. Flotation is provided by the numerous thin, flat, gas-filled chambers. Examine any specimens carefully to determine if they represent a recent death or have been floating shells for some little time before being blown ashore (Jongbloed *et al.*, 2016).

*Scaphopods* (Figs 14.4, 14.5): (Adapted from Donovan, 1990a, b). The main features of scaphopod anatomy are illustrated in Figure 14.4. Soft part anatomy is explained by Donovan (1990a, pp. 1–2). Growth of the tusk-like shell is at the anterior aperture, associated with some thickening of the wall. The posterior aperture is widened by resorption as the

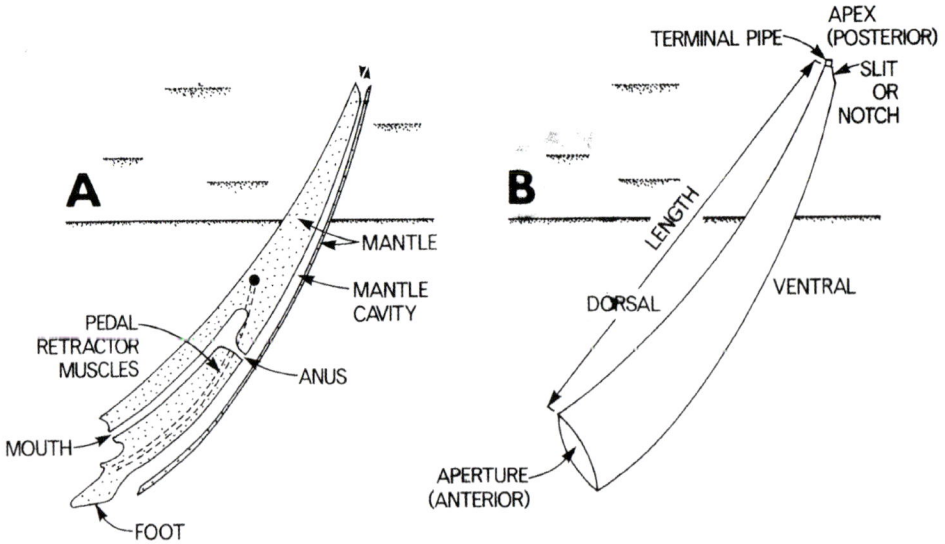

**Figure 14.4** Scaphopods (after Donovan, 1990a, fig. 1). (**A**) Schematic longitudinal section of a living scaphopod, drawn in life position. (**B**) Morphological features of the scaphopod shell.

animal grows. The shell is external, conical, bilaterally symmetrical and gently curved, tapering posteriorly (Fig. 14.4B). In some taxa, the anterior aperture is constricted, so that the greatest shell diameter is just posterior to this region. In other species (Figs 14.4, 14.5), the aperture itself is the broadest part of the shell. The concave side of the valve is dorsal.

**Figure 14.5** Scaphopods (after Donovan *et al.*, 1998, pl. 1, fig. 1). Natural History Museum, London, G11024, general view of a coquinoid slab with abundant scaphopod *Dentalium* sp. and associated gastropods, from the Upper Pliocene Bowden shell bed, Jamaica. The specimen in the centre indicated by an arrow has been bored.

It generally bears fine growth lines perpendicular to the longitudinal axis, marking the positions of previous anterior apertures. Sculpture, where present, is commonly limited to a pattern of anterio-posteriorly orientated ribs, which may form part of a reticulate pattern if the growth lines are particularly prominent. The shell may be transversely circular, elliptical, triangular, tetragonal, octagonal or polygonal, and may be different in different regions of the shell. The posterior aperture may be simple, slitted or notched, with or without a projecting terminal pipe.

I admit to having had little success in collecting modern scaphopods from British beaches (Donovan, in press), but they may form shell accumulations locally, presumably after a die-off and winnowing. They may also be common in the fossil record (e.g., Donovan, 1990a; Donovan & Jagt, 2012). Bruce *et al.* (1963, p. 210) only listed two species from the Irish Sea.

***Chitons*** (Figs 14.6, 14.7): I have a soft spot for chitons for two principal reasons. They are locally common on rocky exposures at Round Hill on the south coast of Jamaica. When I worked at the University of the West Indies, I would take a first-year field trip to this section as one of their first excursions. The geology is excellent, but students were always diverted by the chitons, when I would say a few words about their affinities and poor fossil record. It is the poor fossil record that is my second reason for liking them. Molluscs typically have one or, in bivalves, two valves; chitons have eight. Their poor fossil record is because they

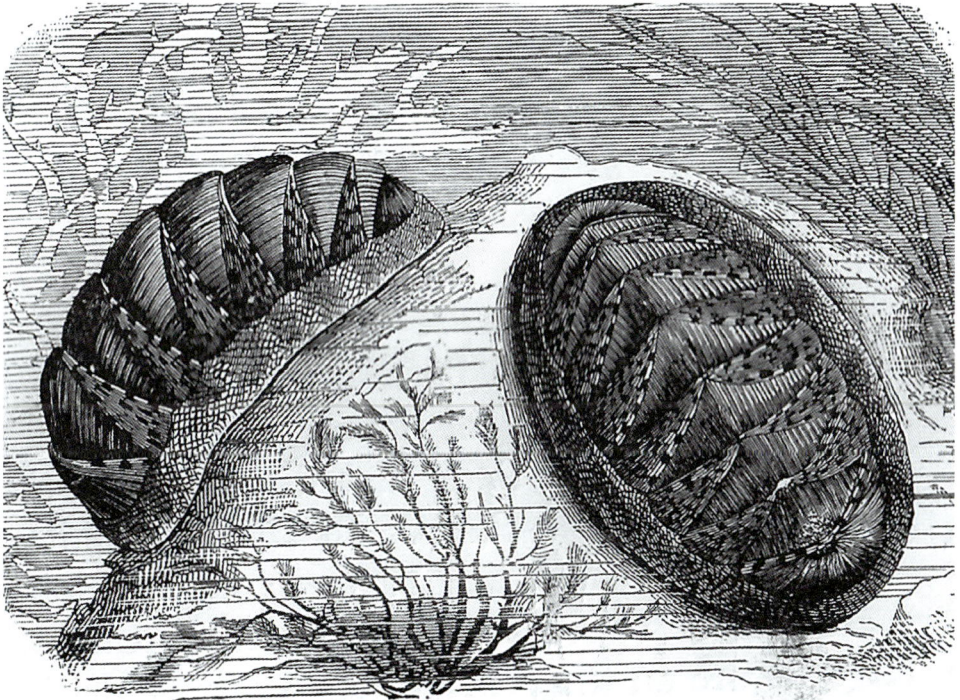

**Figure 14.6** Articulated chiton, West Indian *Chiton squamosus* Linné, in lateral (left) and dorsal (right) views, attached to a lithified substrate (after Step, 1901, fig. on p. 181).

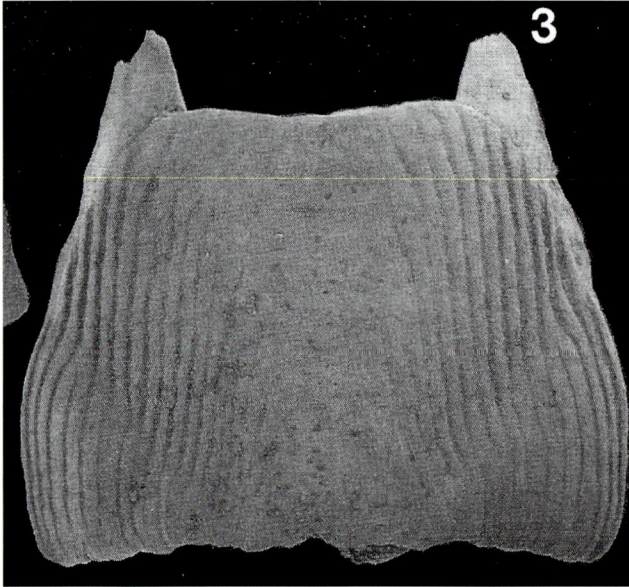

**Figure 14.7** Disarticulated chiton valve (after Donovan *et al.*, 1998, pl. 1, fig. 3). Florida Museum of Natural History, Gainesville, UF 76544, indeterminate chiton valve in dorsal view from the Upper Pliocene Bowden shell bed, Jamaica. Maximum width about 5.5 mm.

disarticulate soon after death and they live on rocky substrates, typically not good sites for fossil preservation.

Of all the molluscs, the chitons have the most archaic appearance with a shell composed of eight overlapping valves (Fig. 14.6). The 'foot' is reminiscent of that of gastropods. The shell commonly has a low, flattened profile befitting an organisms attached to rocks exposed to waves. The separate plates are easily distinguished (Fig. 14.7), but, because chitons are not widely known, their plates are even less familiar to the uninitiated.

## On the beach

The internal shell (= cuttlefish 'bone'; Fig. 14.2) of the common cuttlefish, *Sepia officinalis* Linné, is the cephalopod that you will most likely find on a European beach. On death, the shell will float freely once the soft tissues have rotted away. A storm will probably blow shells onshore. Indeed, the shells are able to float and be carried to beaches well outside the limits of the living cuttlefishes (Donovan *et al.*, 2001).

Bruce *et al.* (1963) listed nine species of chiton and only two of scaphopod from the Irish Sea. Scaphopods may be washed up as a minor component of the allochthonous infauna on British beaches. Chitons will be found living on rocky exposures in the intertidal zone.

## Guides to identification

There are many inexpensive books on the local marine Mollusca. General guides to the seashore life of the British Isles are various, such as Barrett & Yonge (1977), and many newer accounts. Concerning molluscs, my favourite text is Beedham (1972), which covers cephalopods, scaphopods and chitons in adequate detail.

# References

Barrett, J. & Yonge, C.M. (1977) (first published 1958) *Collins Pocket Guide to the Sea Shore*. Collins, London.

Beedham, G.E. (1972) *Identification of the British Mollusca*. Hulton Educational Publications, Amersham.

Bruce, J.R., Colman, J.S. & Jones, N.S. (1963) *Marine Fauna of the Isle of Man and its Surrounding Seas*. Liverpool University Press, Liverpool.

Clarkson, E.N.K. (1998) *Invertebrate Palaeontology and Evolution*. 4th ed. Blackwell Science, Oxford.

Donovan, S.K. (1989) Taphonomic significance of the encrustation of the dead shell of Recent *Spirula spirula* (Linné) (Cephalopoda: Coleoidea) by *Lepas anatifera* Linné (Cirripedia: Thoracia). *Journal of Paleontology*, **63**: 698–702.

———. (1990a) Fossil Scaphopoda (Mollusca) from the Cenozoic of Jamaica. *Journal of the Geological Society of Jamaica*, **27**: 1–9.

———. (1990b) Fossils explained 12: Scaphopods. *Geology Today*, **6**: 128–129.

———. (2000) Cephalopods. In Singer, R. (ed.), *Encyclopedia of Paleontology*: 240–246. Fitzroy Dearborn, Chicago.

———. (in press) Classic localities explained: Beachcombing in Morecambe, Lancashire. *Geology Today*.

Donovan, S.K. & Jagt, J.W.M. (2012) Dentaliids (Mollusca, Scaphopoda) from the type Maastrichtian, the Netherlands and Belgium. In Jagt, J.W.M., Donovan, S.K. & Jagt-Yazykova, E.A. (eds) *Fossils of the type Maastrichtian* (Part 1). *Scripta Geologica, Special Issue*, **8**: 45–81.

Donovan, S.K., Paul, C.R.C. & Littlewood, D.T.J. (1998) A brief review of the benthic Mollusca of the Bowden shell bed, southeast Jamaica. *Contributions to Tertiary and Quaternary Geology*, **35**: 85–93.

Donovan, S.K., Portell, R.W. & Pickerill, R.K. (2001) A shell of the cephalopod *Sepia* Linné from the coast of Carriacou, Grenadines, Lesser Antilles. *Caribbean Journal of Science*, **37**: 125–127.

Jongbloed, C.A., Gier, W. de, Ruiten, D.M. van & Donovan, S.K. (2016) Aktuo-paläontologie of the common cuttlefish, *Sepia officinalis*, an endocochleate cephalopod (Mollusca) in the North Sea. *PalZ*, **90**: 307–313.

Step, E. (1901) *Shell Life: An Introduction to the British Mollusca*. Frederick Warne, London.

# CHAPTER 15

# Worms and their tubes

## Form and function

This chapter looks mainly at two unloved groups of shelly organisms, the serpulid and spirorbid polychaete worms. Invariably attached to a hard (shell, cobble) or seaweed substrate, they are locally common, small and gregarious. These worms are often ignored or receive only minimal descriptions. Herein, I hope to explain something of their interest and why they should not be ignored.

Spirorbids (Fig. 15.1) are small, spiral, sessile polychaetes that are commonly gregarious and cement to a variety of substrates, including shells, rocks and seaweed. They are a common shoreline shell, washed onshore with their substrate. Fossil spirorbids should not be confused with spiral molluscs such as ammonites, which were not cemented. Spirorbids range from the Cretaceous to Recent, but earlier tubes of similar form – 'spirorbiforms' of uncertain affinity – with different shell structures range back to the Ordovician (Taylor & Vinn, 2006). Spirorbids are closely related to serpulids.

Serpulids (Figs 15.2, 15.3) are similar to spirorbids, but are not spiral, except irregularly in some cases. They are commonly gregarious, and their tubes can overgrow and intertwine in close association (Fig. 15.3). They are commonly a little larger than spriorbids, with more prominent growth lines parallel to the aperture, but are still small, and are commonly preserved on protected areas of dead shells, such as on the inner surface of bivalves and within the aperture of gastropods. In these situations, the host shells were obviously dead. Serpulids also grow on the more exposed parts of these shells, when the host mollusc was alive or dead, but are more prone to abrasion in such positions. The earliest serpulids known are Permian. They may densely infest large, discarded borings like *Gastrochaenolites*, which provide a protected environment (Fig. 15.3).

**Figure 15.1** Recent spiral tube worm *Spirorbis spirorbis* (Linnaeus) (after Anon., 1938, p. 14, top right). Diameter measured across the spiral commonly 3.2 mm or less.

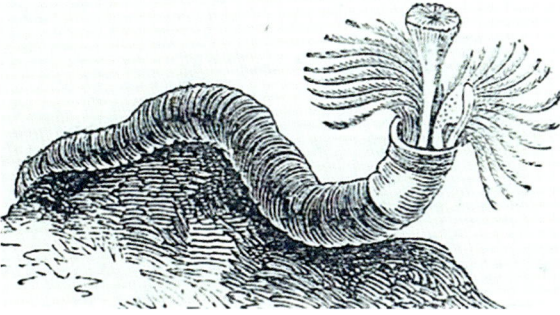

**Figure 15.2** Recent *Serpula* sp. with tentacles expanded (after Taylor, 1885, fig. 137). Scale unknown.

Other 'worms' on the beach are diverse but lack a mineralised portion and are therefore not preserved as dead shells. Some produce agglutinated tubes of sediment grains that may be preserved in borings (e.g., Donovan, 2017a, fig. 3A, B, D). Sabellids are gregarious inhabitants of agglutinated tubes constructed from sediment grains that in some instances form large structures that may be called small reefs. I was first introduced to such structures on the north coast of Jamaica where they form reefs locally but was surprised to find one at the seaward end of a stone breakwater at Cleveleys, north of Blackpool (Donovan, 2021, pp. 200–203).

A further 'hard-shelled' annelid commonly encountered is a boring into limestone clasts and thick bioclasts such as oysters; the substrate acts as the protective 'shell'. *Caulostrepsis* Clarke is a distinctive, U-shaped boring most commonly encountered as *Caulostrepsis taeniola* Clarke (such as Figs 15.4, 17.3B, C, E, H, 18.5A). It is produced mainly by annelids of the family Spionidae (such as *Polydora* spp., common and moderately diverse around the coast of the British Isles; Bruce *et al.*, 1963), but also other polychaete worms (Bromley, 2004,

**Figure 15.3** (After Donovan, 2017a, fig. 3K.) Chalk cobble from the beach between Overstrand and Cromer, north coast of Norfolk, East Anglia, UK (see Chapter 20). Longitudinal section of large boring *Gastrochaenolites* isp. densely coated with the gregarious serpulid *Hydroides norvegica* Gunnerus; a bryozoan encrusts the serpulids (upper right). Specimen uncoated. Scale bar represents 10 mm.

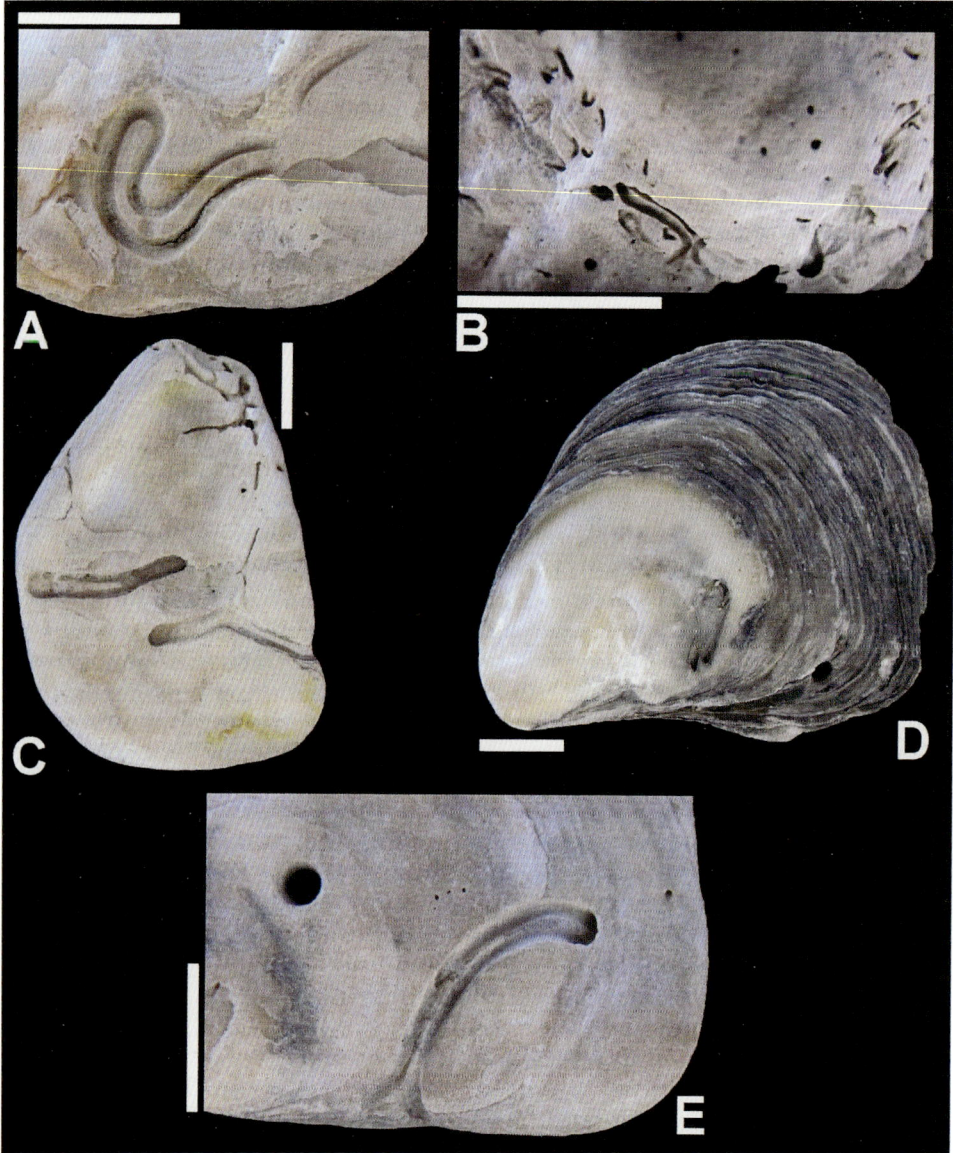

**Figure 15.4** (After Donovan, 2017b, fig. 2). The polychaete worm boring *Caulostrepsis taeniola* Clarke in disarticulated oyster valves from Queen Victoria's bathing beach, Osborne House, East Cowes, Isle of Wight. (**A**) Naturalis Biodiversity Center, Leiden, the Netherlands (prefix RGM) 792 583, a large and highly sinuous specimen partly exposed on inner surface of oyster valve. (**B**) RGM 792 584, several gregarious and smaller specimens on the outer surface of an oyster valve and associated with the sponge boring *Entobia* isp. (= small round holes). (**C**) RGM 792 585, two large, well-exposed borings on inner surface of oyster valve and associated with *Entobia* isp. (upper right). (**D**) RGM 792 586, incompletely exposed specimen (lower centre) on outer surface of oyster valve. (**E**) RGM 792 587, well-exposed and complete large boring exposed on the outer surface of an oyster valve. Note the divergence of the two branches (lower centre) at the apertural end. Specimens uncoated. All scale bars represent 10 mm.

p. 460). This gregarious infestation is most commonly apparent as small, slot-like boreholes, elliptical to figure-of-eight shaped. Commonly, the boring may be exhumed in section, and close examination shows that the shaft is divided by a central, longitudinal vane and is an elongate U-shape (Fig. 15.4).

## On the beach

Serpulids and spirorbids encrust substrates, commonly hard surfaces such as shells and limestone clasts, but not invariably; see the spirorbids on seaweed in Figure 15.1. The same general rules of preservation apply to these worm tubes as to encrusting bryozoans (Chapter 9). Both groups of worm tubes are commonly found in local gregarious accumulations.

Sabellids are uncommon on the beach, although agglutinated tubes may be protected and preserved within borings. The 'reef' at Cleveleys mentioned above was more than 1 m in diameter and I regret not having a camera with me in the field that day. I suspect it was attached to a limestone boulder in the breakwater and may have been buried by sand by the time I returned with my camera. Next time I'm in Cleveleys I shall take a spade, too.

*Caulostrepsis* is common and one of the 'trinity' of borings found commonly around the British Isles and elsewhere (Donovan *et al.*, 2019). It also occurs in different sizes (Donovan, 2017b), either small and gregarious (e.g., Fig. 17.3B, C, E) or large and solitary (Fig. 15.4). This surely reflects infestation by at least two different species of *Polydora*, one small, one big. *Caulostrepsis* is particularly well seen in oyster shells, where the layers of calcite spall off to reveal the structure of the borings within.

## Guides to identification

All the usual suspects. I am not aware of specialist identification guides for these worms, although they doubtless exist, but they are covered, albeit in no great detail, by all the standard guides listed in other chapters.

## References

Anon. (1938) *The Sea-Shore.* W.D. & H.O. Wills, Bristol.

Bromley, R.G. (2004) A stratigraphy of marine bioerosion. In McIlroy, D. (ed.) *The Application of Ichnology to Palaeoenvironmental and Stratigraphic Analysis.* Geological Society, London, Special Publication, **228**, 455–479.

Bruce, J.R., Colman, J.S. & Jones, N.S. (1963) *Marine Fauna of the Isle of Man.* Liverpool University Press, Liverpool.

Donovan, S.K. (2017a) Neoichnology of chalk cobbles from north Norfolk, England: Implications for taphonomy and palaeoecology. *Proceedings of the Geologists' Association*, **128**: 558–563.

———. (2017b) Two forms of the boring *Caulostrepsis taeniola* Clarke on Queen Victoria's bathing beach, East Cowes. *Wight Studies: Proceedings of the Isle of Wight Natural History & Archaeological Society*, **31**: 98–101.

———. (2021) *Hands-On Palaeontology: A Practical Manual.* Dunedin Academic Press, Edinburgh.

Donovan, S.K., with Donovan, P.H. & Donovan, M. (2019) (for 2018) A recurrent trinity of Recent borings in clasts around the southern and western North Sea. *Bulletin of the Geological Society of Norfolk*, **68**: 51–63.

Taylor, J.E. (1885) *Our Common British Fossils and Where to Find Them.* Chatto & Windus, London.

Taylor, P.D. & Vinn, O. (2006) Convergent morphology in small spiral worm tubes ('*Spirorbis*') and its palaeoenvironmental implications. *Journal of the Geological Society, London*, **163**: 225–228.

# CHAPTER 16

# Echinoids

## Form and function

There are five extant classes of echinoderm, the spiny-skinned animals: the crinoids (sea lilies); the echinoids (sea urchins); the asteroids (starfishes or sea stars); the ophiuroids (brittle stars); and the holothuroids (sea cucumbers). Certainly in Europe, the group most likely to be washed up on the beach are the echinoids. The echinoids appeared in the Ordovician but remained a relatively minor group throughout the Palaeozoic. Only two lineages survived the end Permian mass extinction, but the echinoids underwent a major adaptive radiation during the Jurassic. Recent shallow water echinoids are common and diverse. The echinoid test is relatively robust when compared with that of other echinoderms. In consequence, the echinoids have the most complete post-Palaeozoic fossil record. Fossil echinoids are rare as beach clasts at certain sites (see Chapters 17, 20).

The division between regular and irregular echinoids (Figs 16.1, 16.2) is made on the basis of the position of the periproct. This is the opening of the test for the anus, which in life is covered by a periproctal membrane which is usually plated. In regular echinoids the periproct occurs within the apical system (Fig. 16.1), a group of plates at the apex of the test which includes the madreporite. Regular echinoids are generally radially (pentamerally) symmetrical, which is slightly imperfect due the madreporite being slightly enlarged when compared with other ossicles in the same plate circlet (Fig. 16.1; see the upper right of the apical system). In rare examples, such as *Echinometra* Gray, the test is elongated and the symmetry is therefore bilateral, more or less (Fig. 16.3). In irregular echinoids, the periproct is outside the apical system and is said to have moved posteriorly (Fig. 16.2, at bottom of right-hand specimen). This eccentric periproct gives all irregulars an anterio-posterior axis and bilateral symmetry (note that the bilateral symmetry in irregulars is in a different orientation to that of the elliptical regular echinoid *Echinometra*, with respect to the position of the madreporite). All regular echinoids are epifaunal, whereas most irregular echinoids are adapted for an infaunal existence.

The echinoids are characterised by having a globular to flattened, spinose test. The test may be flexible (Fig. 16.4A) or, far more commonly, rigid (Figs 16.1, 16.2, 16.4B–D, 16.5). Adjacent ossicles of the test are held together by mutable collagenous tissues. In many species, the test is further strengthened by the stereom trabeculae of adjacent plates interlocking (Fig. 16.4C). The strongest tests are found in the clypeasteroids (sand dollars),

**Figure 16.1** A Recent regular echinoid, *Eucidaris tribuloides* (Lamarck) (after Donovan & Lewis, 2009, fig. 1C), Naturalis Biodiversity Center, Leiden, the Netherlands (prefix RGM) 554 911, from the Palisadoes, south coast of Jamaica. Apical view. The apical system is central and near-circular. The five large, radially arranged plates at the margin of the apical system are genitals; the large genital plate at 1 o'clock is the madreporite. Note that the genital plates are perforated by prominent genital pores, one per plate. The five genital plates are separated by five smaller, triangular ocular plates, making a circlet of ten plates. Within this circlet are the smaller, more numerous periproctal plates that, in turn, surround the central periproct. Scale in mm.

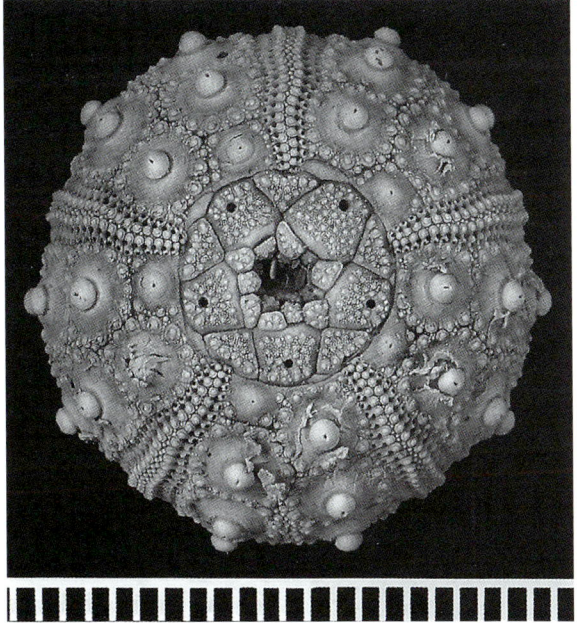

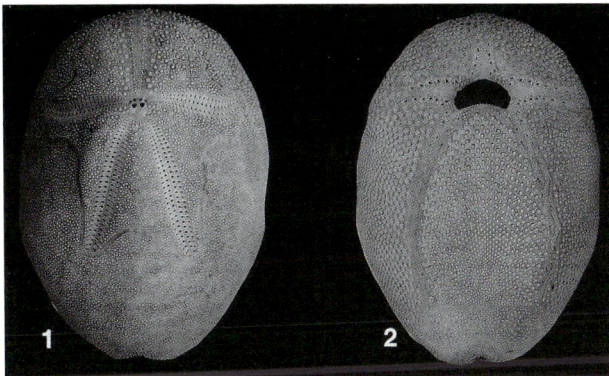

**Figure 16.2** A Recent irregular echinoid, *Brissus unicolor* (Leske) (after Donovan & Veale, 1996, fig. 4.1, 4.2). The Palisadoes, parish of St Andrew, Jamaica (see Chapter 32). Two specimens, each about 43 mm in length, registered at the National Museum of Natural History, Smithsonian Institution, Washington, DC, registration USNM E44054. (**1**) Apical view. Apex towards anterior; ambulacra petaloid apart from that in the anterior; the periproct is posterior, but not apparent in this view. (**2**) Oral view. Peristome kidney-shaped and towards anterior with the pores of the ambulacra easily distinguished; the posterior periproct is just apparent. Contrast the size and density of tubercles with *E. tribuloides* (Fig. 16.1); each tubercle would bear a spine in life. *Eucidaris tribuloides* is epifaunal and the spines are defensive; *B. unicolor* is infaunal and the spines are short, protecting the test from sediment and facilitating digging. Specimens painted with food colouring and whitened with ammonium chloride sublimate.

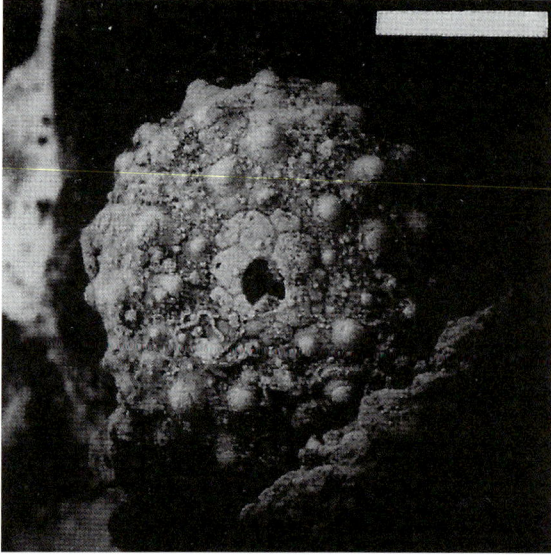

**Figure 16.3** A Late Pleistocene regular echinoid, *Echinometra viridis* A. Agassiz (after Donovan & Collins, 1997, fig. 1), Natural History Museum, London, BMNH EE 5247, from the Falmouth Formation, East Rio Bueno Harbour, parish of St Ann, Jamaica. Note the well-preserved apical system and the elliptical outline of the test. Scale bar represents 5 mm.

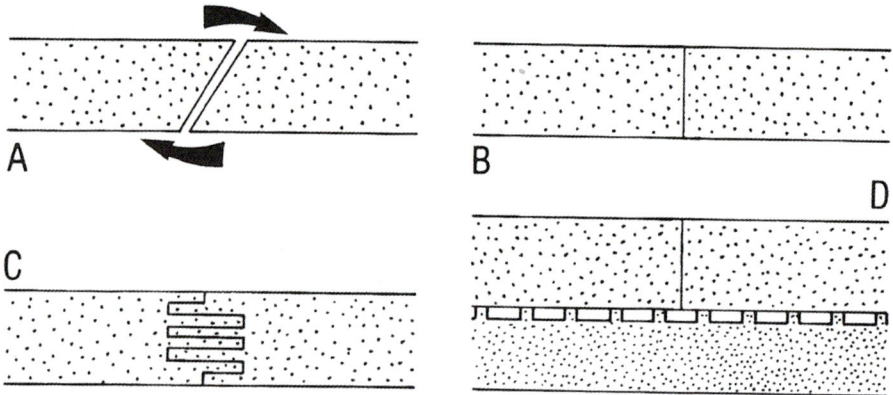

**Figure 16.4** Schematic diagrams of the geometry of plate sutures within the echinoid test (after Donovan, 1991, fig. 11.2). (**A**) A flexible test; arrows indicate relative movement of plates. Such an echinoid would be unlikely to be found on the beach unless it was a reworked Palaeozoic fossil. (**B**) Rigid test, for example, as found in *Eucidaris tribuloides* (Fig. 16.1). (**C**) Rigid test with strongly interlocking trabeculae (= rods of calcite in the microstructure), for example, as found in *Brissus unicolor* (Fig. 16.2). (**D**) Rigid test of *Clypeaster* with double wall (Fig. 16.5).

which are flattened and have a system of internal calcareous supports (Fig. 16.4D). In the genus *Clypeaster* Lamarck, the test is also thickened by a second, internal calcite layer (Fig. 16.5).

The test is divided into five narrow ambulacra (singular, ambulacrum) separated by five interambulacra (Figs 16.1–16.3, 16.5). In post-Palaeozoic echinoids, each ambulacrum and each interambulacrum is comprised of two columns of ossicles; there are thus 20 columns

**Figure 16.5** A Recent irregular echinoid, *Clypeaster rosaceus* (Linné) (after Donovan, 1993, fig. 10.1, 10.2), USNM E40376, Jamaica. Upper left, apical view. Apical system central, five ambulacral petals. Lower right, oral view. Peristome central; periproct posterior. Scale bar represents 10 mm.

of plates within the test. Each plate column is orientated from the apex to the base of the test. The water vascular system is developed internally in the ambulacral regions. Each ambulacral plate is perforated by one or (more usually) two ambulacral pores. These ambulacral pores/pore pairs form part of the water vascular system, each pore supporting one tube foot (Figs 16.1, 16.2, 16.5). In many post-Palaeozoic echinoids, the ambulacral plates have become fused, or compounded, so that each plate bears multiple pore pairs. The pore pair geometry reflects the morphology of the tube feet, which are often specialised for respiration, feeding or attachment (Smith, 1984).

**Figure 16.6** Spine of Recent *Eucidaris tribuloides* (Lamarck) (after Donovan, 1993, fig. 5.1), BMNH E83646b. Recent, Discovery Bay, parish of St Ann, Jamaica. Lateral view of shaft (total length about 14 mm). Scanning electron micrograph of specimen coated with 60 per cent gold-palladium. Scale bar represents 1 mm.

The spines, or radioles, have a muscular attachment to the test and articulate against domed tubercles (Figs 16.1, 16.6). The adoral (oral) surface of an echinoid is in contact with the sediment, opposite the upper aboral (apical) surface. The opening of the test for the mouth is called the peristome (Fig. 16.2, 16.5), which in life is covered by a peristomial membrane; the mouth is central. In regular echinoids the peristome is in the centre of the adoral surface, but it may be central or anterior (that is, it may have moved in the opposite direction to the periproct) in irregulars. In regulars and some irregulars, the mouth has a group of five protruding teeth which form part of a complex feeding structure called Aristotle's lantern. Where the lantern is lost in irregulars, it is either replaced by scraping teeth or teeth may be completely absent.

In regular echinoids the periproct is encircled by the apical system, composed of ten ossicles arranged in two circlets (Figs 16.1, 16.3). Four large genital plates plus the madreporite are interambulacral in position and all bear genital pores. Five smaller ocular plates are ambulacral in position. The evolution of irregular echinoids led to the periproct migrating out of the apical system in an interambulacral position, resulting in the loss of one genital plate. In irregular echinoids, the periproct may be aboral (but eccentric), posterior or adoral (but posterior of the peristome).

Echinoids are spiny. Spines are attached to the test by collagenous tissues and muscles. These unmineralised tissues rot soon after death and the spines drop off. So, it is common for dead sea urchins washed up on the beach to be 'bald', bereft of the spines that protected them in life (Figs 16.1–16.3, 16.5). Spines may be plentiful on a beach (Chapter 3) but are probably tiny or broken or both. The easy way to find them is to take a sample of shell hash, wash and dry it, and examine it under a lens or microscope. Few echinoid spines are as big and obvious as those of cidaroid echinoids (Fig. 16.6).

## On the beach

Looking for recently dead echinoids is always going to be a matter of luck. In the natural environment they are less common than molluscs; the test is multi-plated and thus susceptible to disarticulation once the soft tissues rot away; and larger tests like the heart urchin *Echinocardium cordatum* (Pennant) may be broken up by the beaks of scavenging seagulls. In short, they are the ideal subject for a small and specialised collection. But there are only about ten or so species known from around the British Isles; a holiday to, say, the Caribbean would be likely to be more productive (Hendler *et al.*, 1995).

One echinoid that may be locally common, yet unobtrusive, is the tiny sand dollar *Echinocyamus pusillus* (Müller) (Donovan, 2025). A big specimen is likely to be less than 10 mm in maximum dimension. They are flattened, rounded to oval in outline and no bigger than a shirt button. Collectors should either get on their hands and knees with a hand lens or make a bulk sample of shell hash for careful inspection at home.

Further, shell hash is a good source of echinoid debris, such as spines and plates of the test, providing you do not mind your echinoids coming in 'kit form'! As echinoids fall apart so readily after death, the best source of their remains is in the fragments. Do not look upon picking fragments as a whimsical pursuit, but a source of excellent material; that is, the echinoids may be disarticulated, but the preservation of individual plates can be exquisite.

Such studies are always likely to yield other interesting fragments, such as crustaceans (crabs and barnacles). Processing a poorly lithified sedimentary rock for disarticulated echinoid remains is a common technique for collecting from many palaeontological sites (e.g., Donovan & Paul, 1998).

Another rarity is to watch for dead, floating echinoid tests (Reyment, 1986). These nekroplanktonic tests dried out on a beach and became mummified, before floating away on a high tide and being transported in death in this unexpected manner.

## Guides to identification

All popular palaeontological textbooks will cover the complexities of echinoid morphology in detail, such as Clarkson (1998, pp. 263–285). The specific diversity of extant echinoids around the British Isles is limited and adequately covered by standard field guides, such as Barrett & Yonge (1977). If more detail is required, the standard text on the British echinoderms is Mortensen (1927). Echinoids are lucky to have a marvellous specialist text, Smith (1984), which is both highly informative and readable. Also of interest, but rather different, is McNamara (2011), which is an enthralling account of the folklore of echinoids.

## References

Barrett, J. & Yonge, C.M. (1977) (first published 1958) *Collins Pocket Guide to the Sea Shore*. Collins, London.

Clarkson, E.N.K. (1998) *Invertebrate Palaeontology and Evolution*. 4th ed. Blackwell Science, Oxford.

Donovan, S.K. (1991) The taphonomy of echinoderms: Calcareous multi-element skeletons in the marine environment. In Donovan, S.K. (ed.), *The Processes of Fossilization*. Belhaven Press, London, 241–269.

———. (1993) Jamaican Cenozoic Echinoidea. In Wright, R.M. & Robinson, E. (eds), *Biostratigraphy of Jamaica*. Geological Society of America Memoir, **182**: 371–412.

———. (2025) A beachcomber's field guide to the other side of Doggerland: Zandvoort aan Zee, Noord Holland, The Netherlands. *Bulletin of the Geological Society of Norfolk*, **75**: 45–55.

Donovan, S.K. & Collins, J.S.H. (1997) Unique preservation of an *Echinometra* Gray (Echinodermata: Echinoidea) in the Pleistocene of Jamaica. *Caribbean Journal of Science*, **33**: 123–124.

Donovan, S.K. & Lewis, D.N. (2009) Paleontological implications of multiple genital pores in the apical system of *Eucidaris tribuloides* (Lamarck), recent of Jamaica. *Caribbean Journal of Science*, **45**: 20–24.

Donovan, S.K. & Paul, C.R.C. (1998) Echinoderms of the Pliocene Bowden shell bed, southeast Jamaica. *Contributions to Tertiary and Quaternary Geology*, **35**: 129–146.

Donovan, S.K. & Veale, C. (1996) The irregular echinoids *Echinoneus* Leske and *Brissus* Gray in the Cenozoic of the Antillean region. *Journal of Paleontology*, **70**: 632–640.

Hendler, G., Miller, J.E., Pawson, D.L. & Kier, P.M. (1995) *Sea Stars, Sea Urchins, and Allies: Echinoderms of Florida and the Caribbean*. Smithsonian Institution Press, Washington, DC.

McNamara, K.J. (2011) *The Star-Crossed Stone: The Secret Life, Myths, and History of a Fascinating Fossil*. University of Chicago Press, Chicago.

Mortensen, T. (1927) *Handbook of the Echinoderms of the British Isles*. Oxford University Press, Oxford.

Reyment, R.A. (1986) Nekroplanktonic dispersal of echinoid tests. *Palaeogeography, Palaeoclimatology, Palaeoecology*, **52**: 347–349.

Smith, A.B. (1984) *Echinoid Palaeobiology*. George, Allen & Unwin, London.

# III

# Where to go and what to see

# CHAPTER 17

## Field guide: Margate, Kent

### Preamble

(Adapted after Donovan, 2020.) I have already explained my philosophy of field guides for palaeontology (Donovan, 2021, ch. 44). Briefly, my strong opinion is that the field guide is the most geological of publications, apart from the geological map, and is an important form of communication for all geologists, particularly the amateur. I finish this book with a series of field guides to beaches in the British Isles that show features of interest to the beachcombing palaeontologist.

I have only the most tenuous link with the Chalk at Margate, on the north coast of Kent. I had only been there once, in the mid-1980s, when my brother and I spent a happy day collecting many tests of the echinoid *Echinocorys* from the Chalk cliffs. Traces, ancient or modern, were of little interest to me at that time. That ended my involvement with the Chalk at Margate until the late Dr Fiona Fearnhead showed me a fist-sized chalk clast that she had collected from the beach (Donovan & Fearnhead, 2017). This clast was riddled with borings, namely *Caulostrepsis taeniola* Clarke and *Gastrochaenolites ornatus* Kelly & Bromley. One *G. ornatus* had changed the direction of growth and had the appearance of branching, not a feature commonly associated with *Gastrochaenolites* (Kelly & Bromley, 1984). This was a specimen to warm the cockles of any ichnologist's heart. This specimen was certainly unusual; what else might be found? I determined to make a collecting trip to the beach at Margate.

### How to get there

Almost invariably I include a locality map with any field guide, but finding the site at Margate is so straightforward that I thought it unnecessary. But a relevant Ordnance Survey sheet may be useful; I am still using the old one-inch map of East Kent (sheet 173), picked up second-hand for a few pounds and dating from 1969, but more recent sheets are available.

Margate has ample car parking space, as would be expected in a major seaside town, and the railway station is conveniently situated close to the seafront. Walk to the front near the station and follow the seawall east (that is, turn right), then more or less north-northeast and again east after the pier. Keep walking east along the seawall, passing a disused amusement area on the left. You are now just below Cliftonville and the beach spreads out ahead of you (NGR TR 362 714; Fig. 1.3).

## Locality details

The specimens described below all came from this part of the beach, extending eastwards. This is a collecting trip for modern borings and their borers, and Chalk fossils, particularly echinoids. The latter can be collected from a horizon rich in *Echinocorys* ex. gr. *scutata* (Leske) in the cliff (Peake, 1967; Mortimore *et al.*, 2001), but they also occur as rare, reworked clasts on the beach, most likely derived from Walpole Rocks offshore. From the cliff they may be pristine, but they also turn up reworked, abraded and bored on the beach. But you can collect *Echinocorys* from many sites (e.g., see Chapter 20).

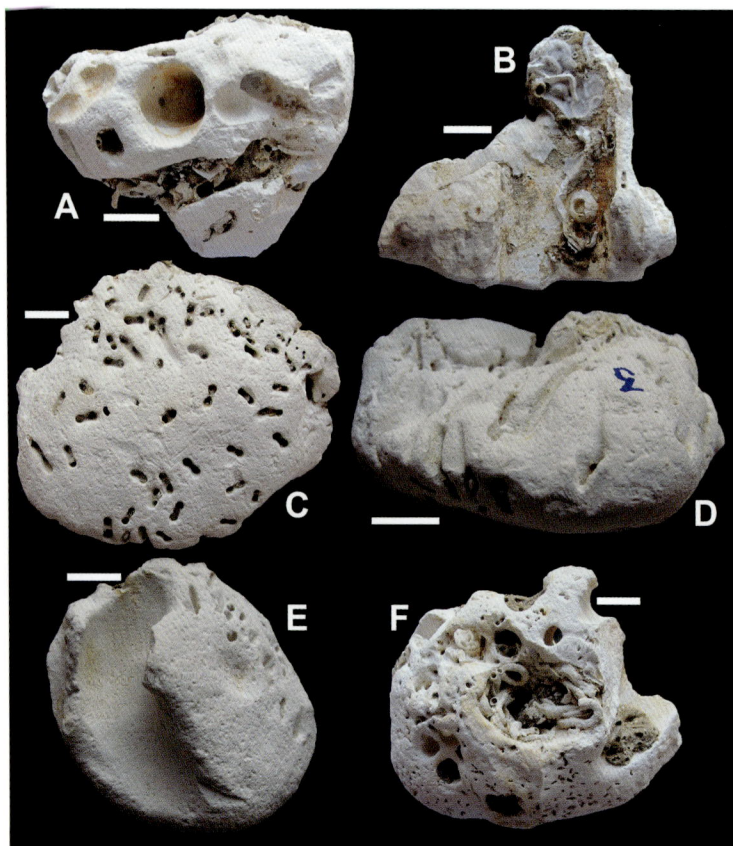

**Figure 17.1** (After Donovan, 2020, fig. 2.) Recent borings in chalk clasts, Margate, Kent. (**A, B**) Two sides of an irregularly shaped cobble, with common *G. ornatus*, which undoubtedly weakened the rock and caused it to break off irregularly shaped fragments as it was moved by the sea. Note that the borings provided protected niches for a variety of encrusting organisms, such as cheilostome bryozoans, serpulid worms and balanid barnacles. (**C, D**) *Caulostrepsis taeniola* in transverse (**C**) and longitudinal section (**D**); the face in (**C**) is towards the bottom of the page in (**D**). (**E**) A curved *G. ornatus* (left) with faint serrations. (**F**) A rounded, but corraded cobble with both *C. taeniola* and *G. ornatus*, the latter providing protective substrates for densely packed encrusting organisms. Specimens uncoated. All scale bars represent 10 mm.

The real attraction of this excursion is to match modern borings with modern borers. As the beach is narrow, it is best to arrive during a falling tide or just after the tide has turned. The rock exposed in the cliffs, and forming clasts on the beach, is Upper Cretaceous chalk which outcrops extensively in this area and extends as Walpole Rocks, the offshore platform. The chalk at Margate is within the Margate Chalk Member, Newhaven Chalk Formation, White Chalk Subgroup, and forms part of the Thanet coast Geological Conservation Review site (Mortimore *et al.*, 2001). All specimens illustrated herein (Figs 17.1–17.4) are deposited in the collections of the Naturalis Biodiversity Center, Leiden, the Netherlands.

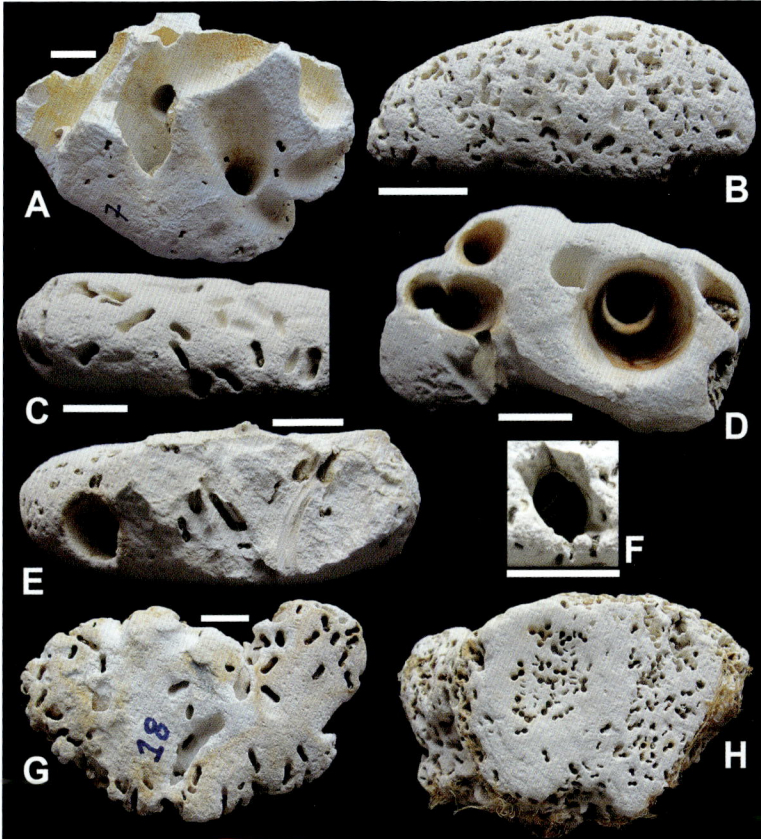

**Figure 17.2** (After Donovan, 2020, fig. 3.) Recent borings in chalk clasts, Margate, Kent. (**A**) Contorted *G. ornatus* borings; the specimen at upper centre shows the sculpture of this species particularly well. (**B**) A rare specimen with both big and small *C. taeniola* in close association. (**C**) Big *C. taeniola* seen mainly in transverse section. (**D**) Several *G. ornatus*. The large boring towards the right, viewed from where its base would have been if it had not been corraded away, gives a good idea of the conical shape of these structures, looking towards the aperture (not preserved). (**E**) Mainly longitudinal and transverse sections of *C. taeniola*, several of which show the central vane particularly well. (**F**) An articulated bivalve in *G. ornatus*, either the producer or more likely a nestler, which took the hole over after it was bored. (**G, H**) The reverse sides of a single slab of chalk, showing the contrast in *C. taeniola* size between the two. Specimens uncoated. All scale bars represent 10 mm.

## What to look for

*The borings*: Of the three ichnotaxa identified in the present study, two are common in Chalk clasts (Figs 17.1, 17.2) and two occur in robust mollusc shells, namely one attached oyster valve, the Portugese oyster *Crassostrea angulata* (Lamarck), and two shells of the common whelk, *Buccinium undatum* (Linné) (Fig. 17.3). The two ichnotaxa mentioned above occur commonly in the chalk clasts of Margate, namely the U-shaped boring

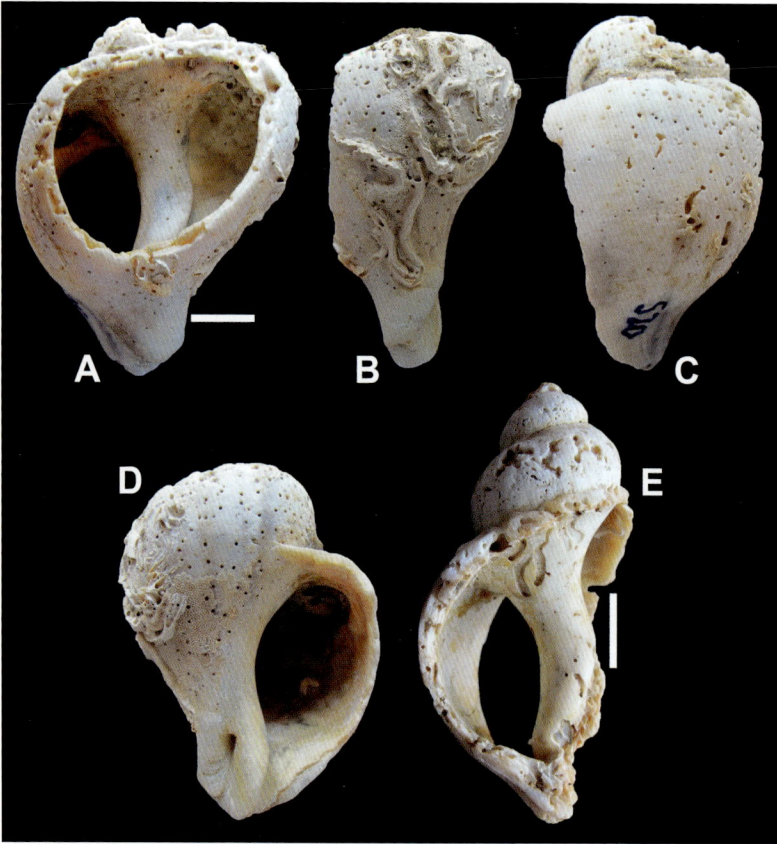

**Figure 17.3** (After Donovan, 2020, fig. 4.) Recent borings in the common whelk, *Buccinium undatum* Linné. (**A–D**) Four lateral views of a single shell. (**A**) Reverse side (abapertural), showing a corraded, rounded hole in part of the shell that was probably weakened by borings. Chambers of *Entobia* isp. are seen in the section through the shell wall; small holes are apertures of the same; balanid barnacles are among the interior denizens encrusting of the shell. Note that this broad hole lies flat, indicating corrasion rather than impact damage. (**B**) Rotated 90° to the left from (**A**). A dense community of *Entobia* isp. (small holes), serpulid worms and cheilostome bryozoans. (**C**) Rotated 90° to the right from (**A**). Part of the shell infested by mainly *Entobia* isp. (**D**) Apertural view with common *Entobia* isp. apertures, plus bryozoans and serpulids, both of which show signs of sponge borings. (**E**) A second shell, but one in which *C. taeniola* is the more prominent boring. Specimens uncoated. Scale bars represent 10 mm.

*Caulostrepsis taeniola* Clarke and the larger, club-shaped boring *Gastrochaenolites ornatus* Kelly & Bromley. A third boring, the complex gallery pattern of *Entobia* isp., is surprisingly rare and limited to robust mollusc valves.

*Caulostrepsis taeniola* is gracile, gregarious and may penetrate deeply into a limey or shelly substrate (Figs 17.1C–E, 17.2A–C, E, G, H, 17.3E). It is U-shaped with a central vane of rock or shell dividing the two branches (see Chapter 15). In cross-section, the boring is shaped like a figure of eight. In the present collection, it is found in chalk clasts, whelks and oysters. The Margate *C. taeniola* falls into two size ranges (such as on the two sides of Fig. 17.2G, H), which may conveniently be called big and small; these tend not to occur together on the same surface, with one exception (Fig. 17.2B), where they probably represent two separate periods of infestation. A similar pattern of infestation by *C. taeniola* has already been described from the Isle of Wight (Donovan, 2017). In the North Sea these borings are produced by spionid polychaete annelid worms, *Polydora* spp. (Donovan, 2017).

*Gastrochaenolites ornatus* are easily the largest borings in the chalk clasts at Margate (Figs 17.1A, B, E, F, 17.2A, D–F), not occurring in any of the shelly substrates, and are invariably incomplete. A complete *G. ornatus* would be a club-shaped pit of circular section throughout; the chamber where the shell of the producing boring bivalve nestled had a subtle, serrated sculpture (Kelly & Bromley, 1984; Figs 17.1E, 17.2A). That some *G. ornatus* are 'old' is suggested by the encrusting worms in some of these borings (Fig. 17.1A, B, F). At least two borings contain small articulated bivalves (Fig. 17.2F), either the producers or nestlers, living in the protection of a discarded boring. The producer of *G. ornatus* is a boring pholadid bivalve (see below).

*Entobia* isp. is rare at Margate (Fig. 17.3). It is certainly there in both whelks and also an oyster valve, but not in the chalk clasts. *Entobia* is a shallow network of borings made by clionaid sponges and is easily recognised from its multiple, small circular apertures that perforate the substrate surface, in close association and more or less regularly spaced. Where broken through, the network of tunnels and chambers is apparent (Fig. 17.4A, E).

**The borers**: Of the three groups of organisms that produced the above suite of borings – spionid polychaetes, clionaid sponges and pholadid bivalves – it is only the last that has a shell with a reasonable preservation potential. The beach in Margate is unique in my experience. Beaches which front a sandy sea bottom are commonly dominated by burrowing bivalves, of which many thousands of disarticulated valves may litter a strandline. It is only rarely that a valve of a boring bivalve is encountered in such settings, certainly only a small fraction of 1 per cent of the specimens.

In stark contrast, all of the infaunal bivalves found on the beach at Margate were borers, the pholadid *Pholas dactylus* Linné (Fig. 17.4). All of the specimens were disarticulated (with one exception) and included an approximately equal mix of right (Fig. 17.4A, B) and left (Fig. 17.4D–F) valves. The exception is a pair of articulated valves kept in close association by the holdfast of a seaweed (Fig. 17.4C), attached to the posterior (= siphonal) end. The valves are thin and commonly broken. It is likely that they are derived from just offshore where they were boring in the chalk platform. It is a reasonable assumption that *P. dactylus* was the producer of the *G. ornatus* borings, described above.

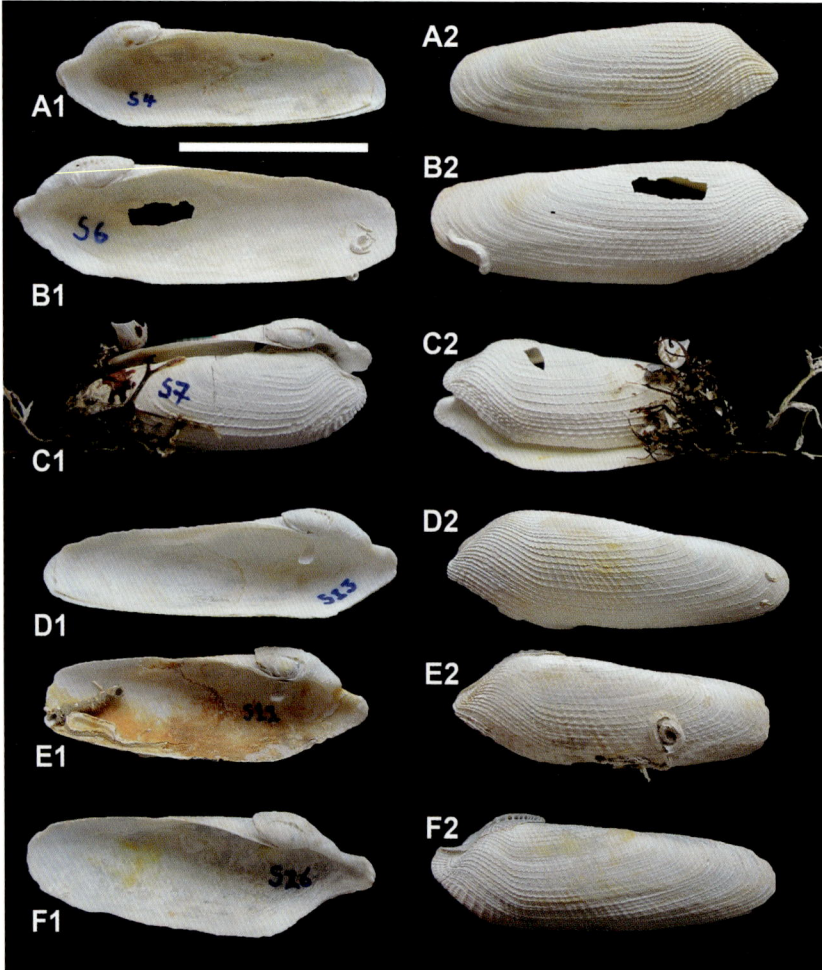

**Figure 17.4** (After Donovan, 2020, fig. 5.) Right valves (**A**, **B**), an articulated shell (**C**) and left valves (**D–F**) of mature *Pholas dactylus* Linné. Note features such as damage (the valves are thin), and worm tubes on the inner and outer surfaces. Further note how much larger these shells are than the borings of *G. ornatus* in Figures 17.1 and 17.2, even though this was the likely producer. (**A, B, D–F**) (**1**) = internal views of valves; (**2**) = exterior views of same valves. (**C**) Articulated shell held together by growth of seaweed at the posterior (siphonal) end; **C1** right valve view; **C2** left valve view. Specimens uncoated. Scale bar represents 50 mm.

## Discussion

The beach at Margate yields bored chalk clasts, whelks and oysters, and valves of *Pholas dactylus*, a notable suite of borers and borings. Yet this is not the end of the story; the observations above lead to a series of questions that stretch our ideas on the ecosystem. These questions and their answers are directly applicable to analogous occurrences in the fossil record.

Why are there two distinctly different sizes of *Caulostrepsis taeniola*? A qualitative inspection of the bored clasts suggests any *C. taeniola* is either big or small with few gradations between them. Further, on most surfaces, all borings are about the same size (but, for an exception, see Fig. 17.2B). At least one specimen (Fig. 17.2G, H) has one surface densely infested by small *C. taeniola* and the opposite side covered by large specimens. The producers are obviously gregarious. These borings may represent two producing species of different sizes or two or more spatfalls of different maturities. The former seems most likely; these borings do not represent a limited time interval but rather were formed at different times during the past tens or hundreds of years, at least. At no time during this period were common borings of intermediate sizes produced. This most likely indicates that there are at least two different species of boring *Polydora*, each producing a different size of domicile; it is significant to note that there are several species of this polychaete known from coastal waters around the British Isles (Bruce *et al.*, 1963).

Why are the *Pholas* valves on the beach bigger than the *Gastrochaenolites* borings? This is probably an indication of the contrasting taphonomic pathways followed by the borers and the bored substrates. Many of the *P. dactylus* valves that were collected were about the maximum size of this species (Fig. 17.4). Because their borings are club-shaped, formed and made progressively wider as the bivalve grows (Fig. 17.2D), it is not possible for a live *P. dactylus* to turn about and emerge from a chalk substrate. After death and rotting of the soft tissues, it may be possible for the separate valves to be washed out of their club-shaped borehole, but is considered unlikely; otherwise, release can only come if the chalk clast is shattered. So, the release of the valves is a destructive event; the big, bored clast is broken. Further, bored clasts are inherently weakened, plus there were not boulders of chalk on the beach; the largest cobble is 110 × 53 × 41 mm, so there is no space for a large *G. ornatus* boring even there. The common incompleteness of these borings is a good indicator that the chalk clasts have been corraded (that is, corroded + abraded). The largest borings of *P. dactylus* are offshore and remain unseen.

Why are there no *Entobia* borings in chalk clasts? It may be that the chalk substrate is unsuitable, yet this is unlikely. *Entobia* is a common boring in chalk clasts on beaches in, for example, north Norfolk and in other limestones elsewhere. Either *Entobia* in chalk is rare at Margate, although this is difficult to explain, or it may have been removed by corrasion of the rims of clasts. Of the three ichnotaxa considered herein, *Entobia* is the shallowest boring. Removal of the thin outer layer of cobbles, bored by a network by sponges, would have accelerated surface corrasion of a rolling clast.

The occurrence of these three ichnotaxa – *Caulostrepsis*, *Gastrochaenolites* and *Entobia* – provides another North Sea site for this trinity of borings (Donovan *et al.*, 2019). In different parts of the southern and western North Sea, the same suite of ichnogenera occur where the coast provides abundant limestone clasts. This is true of Cenozoic limestones (Isle of Wight), Cretaceous chalk (Margate, north Norfolk) and Permian Magnesian Limestone (Easington, Co. Durham). Further, in the Irish Sea, the same borings occur in Carboniferous limestones (Cleveleys, Lancashire). This pattern is regularly recurring and, fortunately, there are many beaches that I have still to visit in pursuit of these data.

# References

Bruce, J.R., Colman, J.S. & Jones, N.S. (1963) *Marine Fauna of the Isle of Man and its Surrounding Seas.* Liverpool University Press, Liverpool.

Donovan, S.K. (2017) Two forms of the boring *Caulostrepsis taeniola* Clarke on Queen Victoria's bathing beach, East Cowes. *Wight Studies: Proceedings of the Isle of Wight Natural History & Archaeological Society,* **31**: 98–101.

———. (2020) Fossils explained 78: Never bored by borings. *Geology Today,* **36**: 232–235.

———. (2021) *Hands-On Palaeontology: A Practical Manual.* Dunedin Academic Press, Edinburgh.

Donovan, S.K., with Donovan, P.H. & Donovan, M. (2019) (for 2018) A recurrent trinity of Recent borings in clasts around the southern and western North Sea. *Bulletin of the Geological Society of Norfolk,* **68**: 51–63.

Donovan, S.K. & Fearnhead, F.E. (2017) A Recent redirected boring, *Gastrochaenolites ornatus* Kelly and Bromley, in the Upper Cretaceous chalk of south-east England. *Bulletin of the Mizunami Fossil Museum,* **43**: 27–29.

Kelly, S.R.A. & Bromley, R.G. (1984) Ichnological nomenclature of clavate borings. *Palaeontology,* **27**: 793–807.

Mortimore, R.N., Wood, C.J. & Gallois, R.W. (2001) *British Upper Cretaceous Stratigraphy* (Geological Conservation Review Series, No. 23). Joint Nature Conservation Committee, Peterborough. https://hub.jncc.gov.uk/assets/92fe7431-a450-46eb-86a8-f2de42904528

Peake, N.B. (1967) The coastal chalk of north-east Thanet. In Pitcher, W.S., Peake, N.B., Carreck, J.N., Kirkaldy, J.F. & Hancock, J.M., *The London Region (South of the Thames).* rev. ed. *Geologists' Association Guides,* **30B**: 1–32.

# CHAPTER 18

# Field guide: Sandown, Isle of Wight

## Preamble

I love family fun on the Isle of Wight, but there is some resentment involved, too. My own parents had a fixation with Cornwall, so for our annual holidays we had to travel down from London mainly on A-roads (my mother did not like motorways) and, once there, there was little to interest me. More than once I suggested that we go to the Isle of Wight instead, but I may as well have asked for the asteroid belt. Ah well. Then I left home for university and my two small brothers were taken on holiday to … the Isle of Wight! And at a later date my mother thought that I had been there, too. If I didn't laugh, I'd cry.

But I did take my own children, Hannah and Pelham, and my girlfriend (now wife), Karen, to the Isle on holiday, and we had a ball. There was so much to fascinate the whole family. Geology was part of the mix, but not to excess. Geologically, I am much of the opinion that the Isle is the sort of place where every square metre is mapped out and the 'property' of some researcher or other. I see no reason to tread on the feet of my fellow geologists, bless them all. So, I gave myself the task of finding interesting collecting for my amusement that would not cause problems with other palaeontologists. Beaches and *Aktuo Paläontologie* formed a happy focus of much of my collecting. I have discussed one excellent beach site in Donovan (2021, pp. 204–209), Queen Victoria's bathing beach near Cowes. Herein, I move east and focus my beachcombing eyes on Sandown and the east coast.

## How to get there

Sandown is one of the major holiday destinations in the east of the Isle. The roads are well-signposted for the motorist and there are numerous buses from Ryde and elsewhere. Sandown railway station is a little way inland, but there is a direct road to the coast. Walk along the beach or seawall north towards the Yaverland Battery. On the way you will pass the Dinosaur Isle Museum (www.dinosaurisle.com), which is well worth a visit.

## Locality details

Guides to the geology of the Isle of Wight are many. Most of those on my bookshelves are a little long in the tooth, such as Norman (1887) and White (1975), but they are still full of

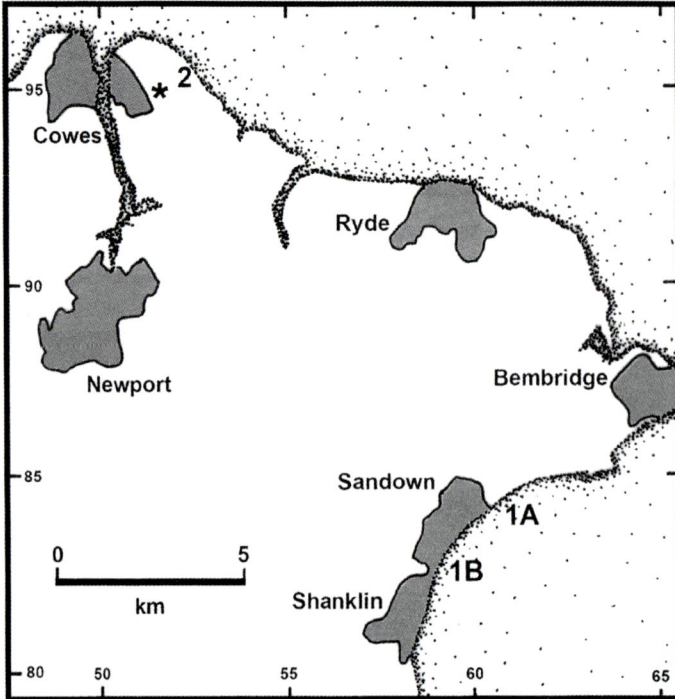

**Figure 18.1** After Donovan (2014, fig. 1). Outline map of eastern and northern Isle of Wight (redrawn and modified after Lloyd & Pevsner, 2006, map on pp. ii–iii). Coastline and sea stippled; principal towns grey. Key: * = Osborne House, Cowes; 1A = Sandown to Yaverland Beach (Fig. 18.2); 1B = Sandown to Shanklin Beach; 2 = Queen Victoria's bathing beach (see Donovan, 2021, pp. 204–209).

**Figure 18.2** After Donovan (2014, fig. 2A). Locality 1A (and 1B in the distance), near Yaverland, and looking towards Sandown and Shanklin. Note that the beach is extensive, has a shallow slope, and is broad and sandy at low tide, with local accumulations of shells, pebbles and cobbles.

meat. The Geologists' Association have published several of their guides to the island. I have yet to buy the latest, but Insole *et al.* (1998) still has much to offer. However, my favourite is my oldest such tome, a rarity, namely *Geological Excursions round the Isle of Wight* [...] by Gideon Mantell (1847).

***Locality 1a: Sandown to Yaverland*** (based on Donovan, 2014, p. 60). The beach between Sandown and Yaverland (Figs 18.1, 18.2) to the east-northeast, in east Isle of Wight [NGR SZ 603 844 to 614 850], is mainly sandy, but there are local accumulations of shells and lithic clasts, including flat slabs of bored mudrock (Fig. 18.3). The ichnofauna (specimens deposited in Naturalis Biodiversity Center, Leiden, the Netherlands (prefix RGM) include common *Caulostrepsis taeniola* Clarke, *Entobia* isp. and *Trypanites* isp., with *Caulostrepsis* isp., *Gastrochaenolites turbinatus* Kelly & Bromley, *Gastrochaenolites* isp. and *Oichnus simplex* Bromley (see below). These borings are found in slabs of mudrock (*Gastrochaenolites* isp. only) and disarticulated oyster valves.

***Locality 1b: Sandown to Shanklin*** (based on Donovan, 2014, p. 60; Fig. 18.1 herein). The beach between Sandown and Shanklin to the south-southwest, in east Isle of Wight [NGR SZ 598 839 to 589 819] (Fig. 18.1), is sandy and has few, sparsely distributed shells. Only one oyster valve was considered worthy of collection (RGM 791 580), bearing *C. taeniola*, *Entobia* isp. and, less certainly, *O. simplex*.

## What to look for

***Mudrock clasts*** (expanded after Donovan, 2014, p. 62) (Fig. 18.3): A moderately common component of the flotsam on the beach at Locality 1A is broad, flat fragments of grey mudrock. The rocks are fissile, reflecting the original bedding. These are derived from offshore and are Cretaceous in origin (Institute of Geological Sciences, 1976); no specimens were found containing fossils and, thus, no more accurate determination could be attempted. Large, round holes penetrate many of these clasts (Fig. 18.3). *Gastrochaenolites* is a club-shaped boring commonly formed perpendicular to the substrate (mudrock, limestone or a large shell). The specimens in Figure 18.3 are interpreted as sections through such large borings perpendicular to bedding.

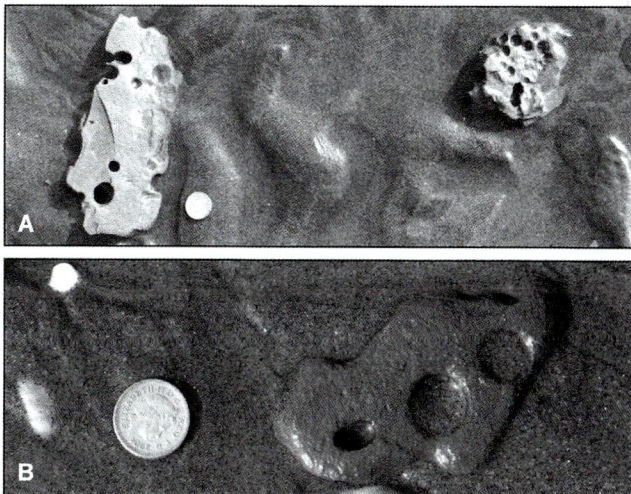

**Figure 18.3** After Donovan (2014, fig. 3). (**A, B**) *Gastrochaenolites* isp. in mudrock slabs at Locality 1A. These large borings are very incomplete and typically consist of sections; the bedding characteristics of the mudrock determined this partial preservation. The right specimen in (**A**) is a moderately thick mudrock cobble that preserves the bases of a number of smaller borings; more typically, slabs are broad and flat. One pound coin for scale (*c*.23 mm in diameter).

Similar occurrences are not uncommon in allochthonous clasts in beach environments and include modern *Gastrochaenolites* borings in Upper Cretaceous chalks of north Norfolk (Donovan, 2011, fig. 2C) and Quaternary peats of the Noord Holland coast, the Netherlands (Donovan, 2013a). Note that the mudrock may be friable and difficult to collect; specimens are better photographed.

***Surface trails*** (adapted after Donovan, 2014, p. 63) (Fig. 18.4): The sandy beaches at Localities 1A and 1B preserve many surface traces and trackways that are produced by intertidal or supratidal vagile invertebrates between successive tides. Some of these can be assigned to well-documented ichnotaxa. But the trail illustrated in Figure 18.4 was particularly distinctive. It is a large trace, continuous over some tens of centimetres and shaped like a W, with an asymmetrical development of markings. There is a slightly off-centre scalloping, flanked by minor 'divoting' on the right, as illustrated. In places there is a shallow groove on the left, as illustrated. The trail makes angular 90° turns with some change in morphology. The direction of movement of the trace maker is uncertain. It is probable that the grooved structure continuous with one end of the trail (Fig. 18.4A, C) is also related.

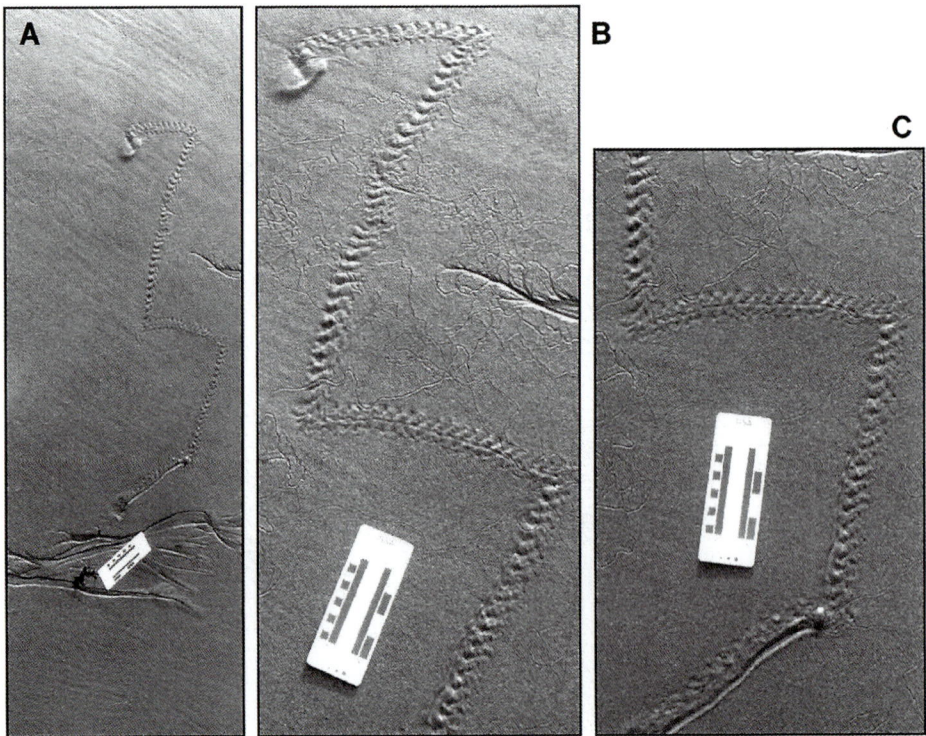

**Figure 18.4** After Donovan (2014, fig. 4). Asymmetrical surface trail, near Yaverland (Locality 1A). (**A**) Complete trace. (**B**) Detail of upper part of trail, showing sharp changes in direction and asymmetry within the trail. The direction of movement is uncertain. Note that some fine, meandering surface trails cut across the asymmetrical trail and thus post-date its formation. (**C**) Lower part of trail; the groove towards 8 o'clock is presumably related to the asymmetrical trail, representing a change of motion of the producer. Scale bar in cm (left) and inches (right).

**Figure 18.5** After Donovan (2014, pl. 1). Borings in the oyster *Ostrea edulis* Linné, all from Locality 1A. (**A, B**) Naturalis Biodiversity Center, Leiden, the Netherlands (prefix RGM) 791 570. (**A**) External surface encrusted by serpulid and spirorbid worm tubes (none on inner surface), with *Caulostrepsis taeniola* Clarke near the umbo. (**B**) Lateral view (umbo towards bottom of page) showing two apertures of different sizes of *Caulostrepsis taeniola*. (**C**) RGM 791 575, external surface of valve strongly infested by *Entobia* isp. (**D**) RGM 791 571, external surface, borings attributed to *Trypanites* isp. in right centre. (**E**) RGM 791 574, external surface, valve showing numerous apertures of *Entobia* isp. (**F**) *Gastrochaenolites* isp. cf. *G. turbinatus* Kelly & Bromley in a thick valve, external surface; borings parallel to plane of paper, that on the left towards 1 o'clock and on the right towards 10 o'clock. All scale bars represent 10 mm.

The asymmetry of the trail suggests a similar asymmetry in the producer, most probably a gastropod. The large size may be due to a predator such as a naticid, but these are typically infaunal burrowers in sand (Frey *et al.*, 1986, figs 2, 3; Morton, 2008), not epifaunal. Although naticids are boring predators, they produce bevelled borings (= *Oichnus paraboloides* Bromley), yet only the distinct *Oichnus simplex* Bromley has been identified from this site.

**Oysters** (adapted after Donovan, 2014; Fig. 18.5): Oysters are rarely pretty shells, but they are a delight for gourmets, shell collectors and palaeontologists for different reasons. The valves of an oyster are large when compared with those of most other common bivalves around the coast of the British Isles. In life, the external surface of an epifaunal oyster may provide a hard substrate for various marine plants and sessile invertebrates (including borers) such as juvenile oysters. Oysters themselves attach to a variety of substrates, commonly firm or hard, but sometimes exotic (Donovan, 2013b). Following death and disarticulation of the valves, both the inner and upper surfaces may be encrusted, and also bored more intensely. A disarticulated oyster valve, replete with encrusting and boring organisms, may thus tell a number of ecological tales to an informed observer.

Oyster valves, attributed to *Ostrea edulis* Linné, may be moderately common at Locality 1A and rarer at 1B. These valves bear some encrusting shells, but, more commonly, preserve a diversity of borings produced by various invertebrates. Borings in shells and pebbles weaken the substrate, and lead to patterns of breakage during corrosion and abrasion that produce modifications in recognisable patterns.

Borings in oysters include *Caulostrepsis taeniola* Clarke (1A and 1B), *Entobia* isp. (1A and 1B), *Gastrochaenolites turbinatus* Kelly & Bromley (1A only), *Oichnus simplex* Bromley (1A and 1B) and *Trypanites* isp. (1A only). The *Caulostrepsis – Entobia – Gastrochaenolites* association is a trinity of borings common off England's east coast (Donovan *et al.*, 2019), preserved on beaches as allochthonous elements of the *Trypanites* ichnofacies.

These borings were made by annelids (*Caulostrepsis*), 'worms' *sensu lato* (*Trypanites*), sponges (*Entobia*), bivalve molluscs (*Gastrochaenolites*) and an indeterminate producer(s) (*Oichnus*). Further, rare shells preserve seagull beak marks. The common occurrence of borings on the internal surface of oyster valves indicates post-mortem infestation.

## References

Donovan, S.K. (2011) Aspects of ichnology of Chalk and sandstone clasts from the beach at Overstrand, north Norfolk. *Bulletin of the Geological Society of Norfolk*, **60** (for 2010): 37–45.

———. (2013a) A distinctive bioglyph and its producer: Recent *Gastrochaenolites* Leymerie in a peat pebble, North Sea coast of the Netherlands. *Ichnos*, **20**: 109–111.

———. (2013b) An unusual association of a Recent oyster and a slipper limpet. *Deposits*, **35**: 5.

———. (2014) Bored oysters and other organism-substrate interactions on two beaches on the Isle of Wight. *Wight Studies: Proceedings of the Isle of Wight Natural History & Archaeological Society*, **28**: 59–74.

———. (2021) *Hands-On Palaeontology: A Practical Manual*. Dunedin Academic Press, Edinburgh.

Donovan, S.K., with Donovan, P.H. & Donovan, M. (2019) (for 2018) A recurrent trinity of Recent borings in clasts around the southern and western North Sea. *Bulletin of the Geological Society of Norfolk*, **68**: 51–63.

Frey, R.W., Howard, J.D. & Hong, J.-S. (1986) Naticid gastropods may kill solenoid bivalves without boring: Ichnological and taphonomic consequences. *Palaios*, **1**: 610–612.

Insole, A., Daley, B. & Gale, A. (1998) The Isle of Wight. *Geologists' Association Guides*, **60**: v + 132 pp.

Institute of Geological Sciences. (1976) *Isle of Wight. England and Wales Special Sheet. Solid and Drift Edition. 1:50,000 Series.* Ordnance Survey, Southampton.

Lloyd, D.W. & Pevsner, N. (2006) *The Buildings of England. The Isle of Wight.* Yale University Press, New Haven, CT.

Mantell, G.A. (1847) *Geological Excursions Round the Isle of Wight and along the Adjacent Coast of Dorsetshire; Illustrative of the Most Interesting Geological Phenomena, and Organic Remains.* Bohn, London.

Morton, B. (2008) Biology of the swash-riding moon snail *Polinices incei* (Gastropoda: Naticidae) predating the pipi, *Donax deltoides* (Bivalvia: Donacidae), on wave-exposed sandy beaches of North Stradbrooke Island, Queensland, Australia. *Memoirs of the Queensland Museum. Nature,* **54**: 303–322.

Norman, M.W. (1887) *A Popular Guide to the Geology of the Isle of Wight.* Knight's Library, Ventnor.

White, H.J.O. (1975) *A Short Account of the Geology of the Isle of Wight.* 3rd impr. [first published 1921]. HMSO, London.

# CHAPTER 19

# Field guide: Walton-on-the Naze, Essex

## Preamble

After my first fossil collecting foray, to the Gault Clay Formation at Folkestone, Kent, only the second stratigraphic unit from which I collected fossils was the Plio-Pleistocene Red Crag Formation at Walton-on-the-Naze. I knew of the Folkestone locality by word of mouth from my school friend Paul White, but how did I discover Walton? The Saturday after my first trip to Folkestone I visited the late-lamented Geological Museum in London and, most likely, invested in a copy of *Directory of British Fossiliferous Localities* (Arkell *et al.*, 1954) from the bookshop. I read this avidly. It had this to report about Walton:

> Pliocene, Red Crag (Waltonian) and basal Nodule Bed (exposed after cliff-falls), resting on Eocene, London Clay. *Neptunea contraria* and other gastropods very numerous, entire but waterworn. Lamellibranchs (including *Glycymeris glycymeris* in prominent bands) common. Derived fossils, sharks' teeth, worn bones, etc. in Nodule Bed. (Arkell *et al.*, 1954, p. 33) [Note that the named molluscs have been reclassified as *Neptunea angulata* and *Glycymeris variabilis*; Todd & Parfitt, 2017.]

This was a huge carrot to this nascent collector. Walton, like Folkestone, is easily reached by rail from my then home in north London. Soon, I had joined the Geologists' Association and further supplemented my Walton library with Markham (1973). Walton was a favourite collecting site for Plio-Pleistocene fossils before I left for university in 1977.

Writing this chapter gave me the excuse to pay a nostalgic trip to Walton in May 2023, 48 years after my first collecting visit. Of course, things change, and I had been warned of the deleterious effect on the exposure by the new sea defences by Mercer & Mercer (2022). But rather than lament the passing of 'the good old days', I travelled to assess how the modern Walton could contribute to our present study.

## How to get there

(Adapted from Donovan & Donovan, 1989, p. 57; Fig. 19.1 herein.) To reach this locality by car, take the A133 from Colchester and turn onto the B1033 at Weeley. In Walton, drive

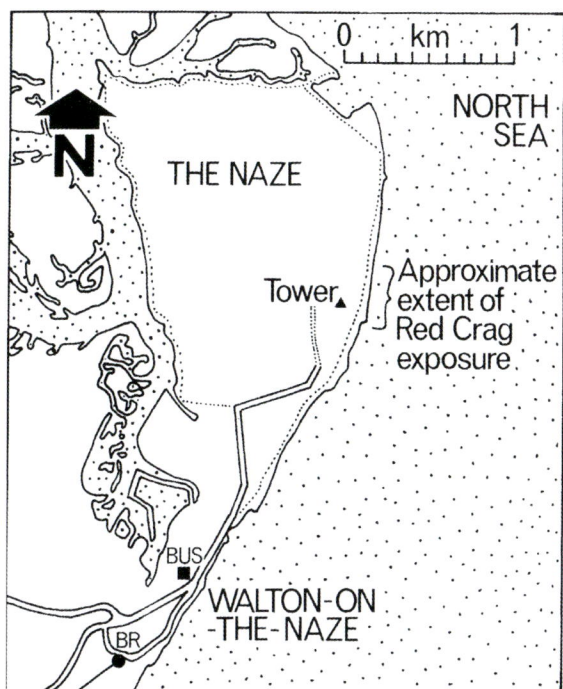

**Figure 19.1** Locality map to show the area of coast where the Red Crag Formation is exposed at Walton-on-the-Naze, Essex (after Donovan & Donovan, 1989, fig. 1). The railway station is marked (BR), and principal roads and footpaths indicated. The most productive beach collecting is on the east coast of the Naze, north from the Tower.

north-northeast along the coast road. The Naze forms a peninsular to the north of this point, marked by the Tower (Figs 19.1, 19.2). The road takes a slight detour away from the coast before returning near a small park. You may leave your vehicle here and continue on foot along the coast path. Alternatively, drive on to the large car park at the Tower and descend by the stairs. If travelling by rail, the same point is reached by walking down the hill from the station to the seafront and continuing along the coast for about 2 km. (The #97

**Figure 19.2** Walton-on-the-Naze, Essex. The old collecting site. The cliffs on the Naze, overshadowed by the tower. The coastal exposure of the Red Crag Formation, near the top of the cliff, and the underlying London Clay Formation, are now protected from the sea by a new sea wall from which this image was taken. The result of reduced erosion is the rocks are becoming overgrown.

**Figure 19.3** Walton-on-the-Naze, Essex. View at beach level, north of the new sea defences. The cliffs and beach exposure are London Clay Formation. Structures in the sea are Second World War pillboxes, which were constructed on the cliffs, but which have dropped to beach level after 80+ years of coastal erosion.

bus from outside the church can save your legs.) The exposure of the Red Crag Formation is reached by descending the slope to the south-east of the Tower and walking along the slipped material about 4 m below the top of the cliff (Fig. 19.2). To visit the beach further north (Fig. 19.3), descend from the cliff top and walk to the new sea defences, with a path on top, and turn left (= north), descending by the stairs at the end.

## Locality details

(Adopted from Donovan & Donovan, 1989, pp. 55, 57.) One of the best deposits in England from which to collect abundant and diverse Plio-Pleistocene shelly faunas is the Red Crag Formation of East Anglia. A small outlier at Walton-on-the-Naze on the Essex coast was formerly well-exposed in a cliff section (Figs 19.1, 19.2) and is a Site of Special Scientific Interest, albeit currently disappearing under a growth of bushes. The Red Crag Formation is exposed as a red-orange shell gravel in the upper part of the cliff at this point, overlain by younger, largely unfossiliferous deposits. I suspect that, with a robust pair of boots and a trowel, good collections can still be made from this section – good hunting. A few gardening tools to hack away vegetation might also be useful.

The underlying grey clay that forms the slipped material is the Eocene London Clay Formation, Walton Member. This is exposed as a line of low cliffs behind the beach (Fig. 19.3) and can also be examined in the foreshore at low tide. The Red Crag Formation rests unconformably on the London Clay Formation.

## What to look for

The beach at the Naze, north from the tower steps (Fig. 19.3), is a delightful mish-mash of Recent shells mixed in with fossils reworked from the Eocene London Clay Formation and Plio-Pleistocene Red Crag Formation. I arrived on a grey day with the tide rising, but in even a relatively brief time I made an interesting collection with camera (Figs 19.4) and collecting bags. The new sea wall in front of the cliffs topped by crag is slowing erosion and favouring the extensive growth of bushes – an enemy of the field geologist – but there is still much of interest to be found north of the sea wall on the beach.

*Pyritised wood*: Among the commonest fossils found in the London Clay Formation at Walton are pieces of pyritised wood, which wash out to form concentrations on the beach and foreshore (Fig. 19.4). These vary from short twigs to hefty branches, made all the heavier by being pyritised. Specimens of wood that were bored by teredinid bivalves as they floated in the London Clay sea are known (Lee *et al.*, 2015, pl. 14d). Pyrite from this beach, called 'copperas', was formerly collected for the manufacture of sulphuric acid (Markham, 1973, p. 10). Other London Clay Formation plant fossils and sharks' teeth are reported to be present, plus rarer invertebrates, although I have had little luck in collecting such Paleogene fossils.

**Figure 19.4** Walton-on-the-Naze, Essex. Pyritised wood in clasts eroded out of the London Clay Formation. The head of the walking stick is about 145 mm in maximum dimension. Note the sandy beach with pebbles of flints and London Clay.

***Shells, modern and ancient***: The best collecting on the Naze was in the Red Crag Formation, where specimens were plentiful and diverse (Arkell *et al.*, 1954, p. 33; Markham, 1973, pp. 7–10). Formerly, loose specimens could be collected washed onto the beach – Red Crag shells are commonly iron-stained orange to brown – but this resource, provided by this Site of Special Scientific Interest (SSSI), has been reduced due to modern coastal defence works (Mercer & Mercer, 2022, pp. 305–309, 390). Not for the first time, a location of international palaeontological significance has been entrusted to the barbarians (compare with Donovan, 2011).

Nevertheless, fossils from the Red Crag Formation continue to form part of the beachload to the north of the SSSI (Fig. 19.3). Common reworked clasts include valves of *Glycymeris variabilis* (Fig. 19.5C) and gastropods such as *Neptunea angulata* (Fig. 19.6A), although any of the illustrated shells might be expected (Figs 19.5, 19.6; see also Todd & Parfitt, 2017). One specimen that particularly intrigued me was a reworked *Glycymeris*, with fresh, likely Recent borings by annelids, *Caulostrepsis taeniola* Clarke (Fig. 19.7A). That is, the shelly substrate and the borings were separated by about two million years. In contrast, a *Neptunea* was the bearer of sponge borings, *Entobia* isp., which may or may not have been coeval.

Reworked shells from the Red Crag Formation are mixed with Recent molluscs on the beach. The latter are easy to separate, being white and certainly fresher in appearance. The most notable specimens on my latest excursion were a pair of cuttlefish 'bones', *Sepia officinalis* Linné (Fig. 19.8). Cuttlefish are a common component of the floating 'nekrobiota' in the North Sea and, after a storm, may litter a beach (Jongbloed *et al.*, 2016).

***Borers and borings***: This beach provides support for the idea of a modern 'boring Trinity' derived from the *Trypanites* ichnofacies offshore (Donovan *et al.*, 2019) around the North Sea coast. *Caulostrepsis taeniola* and *Entobia* isp. are mentioned above, infesting shells.

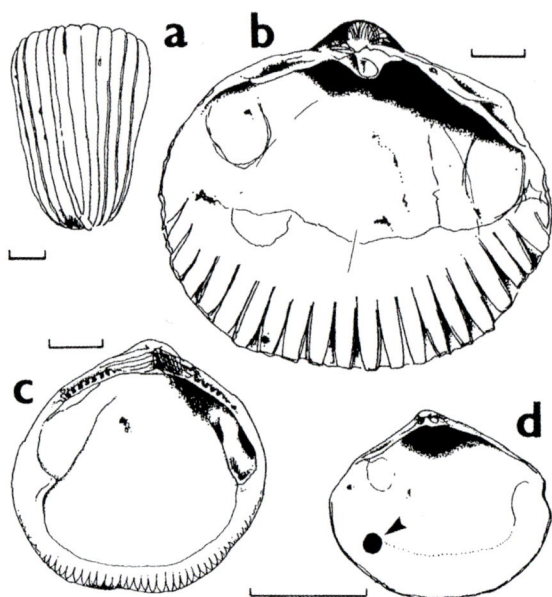

**Figure 19.5** Walton-on-the-Naze, Essex. A solitary coral (**a**) and the internal surfaces of disarticulated valves of bivalves (**b–d**) from the Red Crag Formation (after Donovan & Donovan, 1989, fig. 3; drawn by Mark Donovan from specimens in his collection). (**a**) Lateral view of *Sphenotrochus intermedius*? (Münster); scale bar represents 1 mm. (**b**) Right valve of *Cerastoderma parkinsoni* (J. Sowerby). (**c**) Left valve of *Glycymeris variabilis* (J. de C. Sowerby). (**d**) Right valve of *Digitariopsis obliquata* J. Sowerby; arrow indicates a predatory borehole, *Oichnus* isp. Scale bars in (**b–d**) represent 10 mm. Identifications revised by reference to Todd & Parfitt (2017).

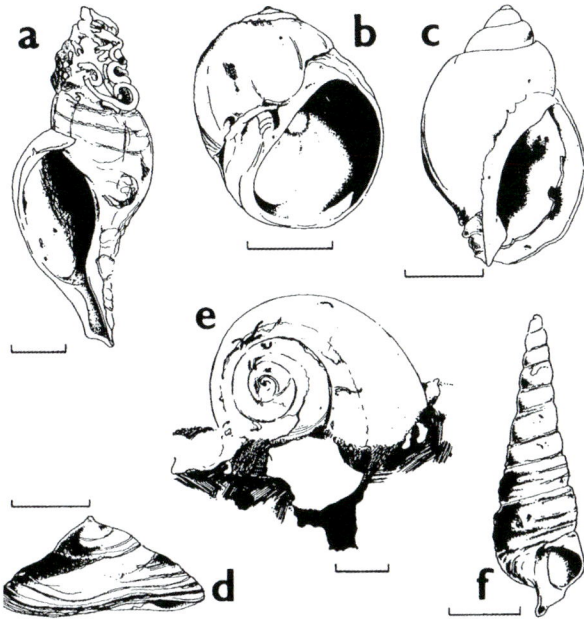

**Figure 19.6** Walton-on-the Naze, Essex. Gastropods from the Red Crag Formation (after Donovan & Donovan, 1989, fig. 2; drawn by Mark Donovan from specimens in his collection). All drawn in apertural view apart from (**d**) and (**e**). (**a**) *Neptunea angulata* Harmer encrusted by serpulid worm tubes. (**b**) *Natica crassa* Nyst. (**c**) *Leiomesus dalei* (J. Sowerby). (**d**) Lateral view of the Chinaman's Hat Limpet, *Calyptraea chinensis* (Linné). (**e**) Apical view of *Euspira catenoides* (S.V. Wood). (**f**) *Potamides tricinctus* (Brocchi). All scale bars represent 10 mm. Identifications revised by reference to Todd & Parfitt (2017).

**Figure 19.7** Walton-on-the-Naze, Essex. Specimens in the collection of the NHMM. (**A**) Reworked right valve of *Glycymeris variabilis* (J. de C. Sowerby), NHMM 2023 046, Red Crag Formation, from the beach and bored by Recent spionid polychaete worms, the trace *Caulostrepsis taeniola* Clarke. (**B**) Cobble, London Clay Formation, NHMM 2023 047, preserving incomplete, multiple *Gastrochaeolites* isp., all borings originating from this side (that is, they expand downwards). (**C**) Recent venerid *Petricola pholadiformis* Lamarck, NHMM 2023 048, right valve, external view. Scale in cm and mm.

**Figure 19.8** Walton-on-the Naze, Essex. Shells of the cuttlefish (Cephalopoda) *Sepia officinalis* Linné washed up at the top of the beach. The head of the walking stick is about 145 mm in maximum dimension. Note that clasts of London Clay dominate (contrast with Fig. 19.4).

The larger *Gastrochaenolites* is particularly obvious in clasts of London Clay, as are the disarticulated valves of bivalve borers (compare with Margate; Chapter 17).

*Gastrochaenolites* are all incomplete because the clasts on the seafloor have been abraded away due to wave transport (Fig. 19.7B). Borings in the London Clay, where present, were all made from one side of a clast, suggesting that these were ripped up fragments of bored seafloor. There is insufficient evidence to classify them beyond *Gastrochaenolites* isp., or, in one instance, *G.* isp. cf. *G. ornatus* Kelly & Bromley. Further, a carbonate nodule was densely infested by small *Gastrochaenolites* isp.

The beach yielded disarticulated valves of two species of boring bivalve, the common pholadid *Pholas dactylus* Linné (again, see Chapter 17) and the rarer venerid *Petricola pholadiformis* Lamarck (Fig. 19.7C), the American piddock. Whether these bivalves produce similar or disparate borings awaits the discovery of better preserved *Gastrochaenolites*, preferably with the borers preserved *in situ*.

## References

Arkell, W.J. and 71 others. (1954) *Directory of British Fossiliferous Localities*. Palaeontographical Society, London.

Donovan, M. & Donovan, S.K. (1989) Site wise: Walton-on-theNaze. *Fossil Forum*, **3**: 55–61.

Donovan, S.K. (2011) Salthill Quarry, Clitheroe: A resource degraded. *Deposits*, **25**: 46–47.

Donovan, S.K., with Donovan, P.H. & Donovan, M. (2019) (for 2018) A recurrent trinity of Recent borings in clasts around the southern and western North Sea. *Bulletin of the Geological Society of Norfolk*, **68**: 51–63.

Jongbloed, C.A., Gier, W. de, Ruiten, D.M. van & Donovan, S.K. (2016) Aktuo-paläontologie of the common cuttlefish, *Sepia officinalis*, an endocochleate cephalopod (Mollusca) in the North Sea. *PalZ*, **90**: 307–313.

Lee, J.R., Woods, M.A. & Moorlock, B.S.P. (eds). (2015) *British Regional Geology: East Anglia*. 5th ed. British Geological Survey, Nottingham.

Markham, R. (1973) Itinerary I: Suffolk and east Essex. In Greensmith, J.T., Blezard, R.G., Bristow, C.R., Markham, R. & Tucker, E.V., The estuarine region of Suffolk and Essex. *Geologists' Association Guides*, **12**: 2–11.

Mercer, I. & Mercer, R. (2022) *Essex Rock: Geology Beneath the Landscape*. Pelagic Publishing, London.

Todd, J. & Parfitt, S. (2017) *British Cenozoic Fossils*. 6th ed. Natural History Museum, London.

# CHAPTER 20

# Field guide: Overstrand and Cromer, Norfolk

## Preamble

I make no excuses for returning to one of my favourite sites in the British Isles, where there is always something intriguing to collect. I presented a field guide to this coast in Donovan (2021a) and cannot resist the chance to, again, walk this productive ground. I try to go to Norfolk for a few days every year, and there is always much to see and collect. Walking from Overstrand to Cromer on a falling tide in the morning makes me feel more alive, with the cliffs bordering the beach, the North Sea disappearing to the horizon and the edifice that is Cromer Pier slowly growing in size as I get nearer (Fig. 20.1). And there are many clasts to examine (Fig. 20.2).

There are, of course, excellent reasons to go over old ground that we feel we know well. As I have emphasised more than once herein, the beach is a dynamic environment. Ordinary processes, day in, day out, shuffle the mix, let alone the influence of a major storm or landslide. What you saw last week, last month or last year will not be the same as you see today. Further, it is the Overstrand to Cromer beach that always emphasises to me that a strand needs to be walked in more than one direction. I walk west in the morning (Overstrand to Cromer) and east after lunch. The tide will be rising or, preferably, falling, either forcing you further up the beach or giving you room to drop down. You will need to be observant in both directions or risk missing that one marvellous specimen that should be in your collection. And recognise that you are always learning new facts and ideas from your beachcombing, enabling you to identify significant specimens that you may have ignored previously.

## How to get there

(After Donovan, 2021a, p. 196.) Take the stopping train from Norwich to Cromer, which runs about once per hour. London to Norwich is quick and easy, but trains from elsewhere can take an age. I prefer to stay in Overstrand, which is a short ride by bus or taxi from Cromer station. Once I am installed, fieldwork is mainly on foot (Figs 20.1, 20.2), but it is useful to know something of the local buses and taxis.

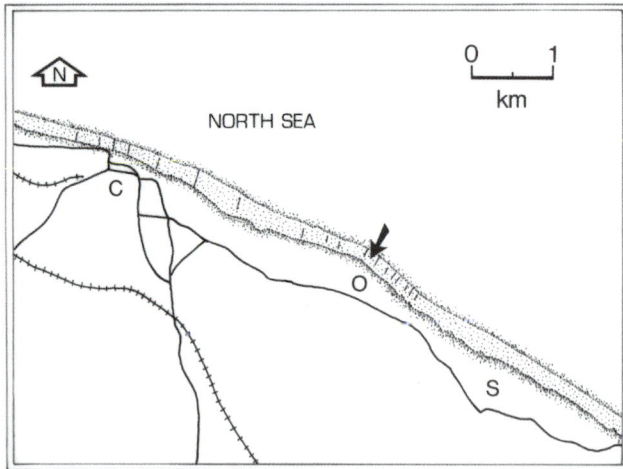

**Figure 20.1** Outline map of the north coast of Norfolk between Cromer (C), Overstrand (O) and Sidestrand (S) (after Donovan, 2010, fig. 1). The dark arrow indicates the author's point of access to the beach. The stippled area is between the low water mark and cliff top; it includes both the beach (groynes are indicated) and slope of the cliffs. Principal roads are shown as solid lines; railways are shown as trellised lines.

## Geological history

See Donovan (2021a, p. 198). The beach at Overstrand is about NGR TH 249 410; that at Cromer, east of the pier, is about NGR TH 227 420 (Fig. 20.1). Although dominantly sandy, the beach also has very numerous Upper Cretaceous pebbles and cobbles, including cobbles of flint and, less commonly, chalk. These were most likely derived from offshore, as at Margate (Chapter 17).

## What to look for

If you refer to Donovan (2021a), you will see that in the equivalent section, the subheadings are 'Chalk fossils', 'Recent borings' and 'Erratics'. While recognising that some overlap is inevitable, I have endeavoured to show more of the diversity of this site than just recycling what is already available. I have two main research interests on this beach, reworked fossils and Recent borings in chalk clasts. Useful sources of information about local Upper Cretaceous fossils and beach erratics are Ewin (2018) and the displays of the Cromer Museum.

*Chalk echinoderms*: Chalk is exposed offshore on the sea bed and as glaciotectonic rafts emplaced in the cliffs. The beach is sandy with many flint clasts and rather fewer cobbles of chalk; the softer chalk is rapidly ground down by the flints under the control of waves.

Fossils occur in both flints and chalk. Flints include common sponges (see below) and other rarer (or less obvious) macrofossils, such as *Steinkerns* of the Upper Cretaceous holasteroid echinoid *Echinocorys* Leske. In the chalk, sponges are less obvious but are certainly there. Other fossils are rare but include *Echinocorys* and other echinoderms. A walk on the beach – about 3 km from Overstrand to Cromer – and return commonly yields at least one Upper Cretaceous echinoid in whatever state of preservation. For example, tests of *Echinocorys* in chalk are rarely complete; they may be broken or have Recent borings, such as *Entobia* isp. and *Gastrochaenolites* isp. (Donovan & Lewis, 2011; Donovan, 2012).

**Figure 20.2** View from the beach between Overstrand and Cromer, north Norfolk, looking west towards Cromer and the pier (after Donovan, 2022a, fig. 2). Note the very numerous cobbles and pebbles of chalk and, mainly, flint on the beach.

Until recently, I had never found a *Micraster* (Donovan, 2022a; Fig. 20.3 right herein) in over twelve years of collecting between Overstrand and Cromer. I always found one or more *Echinocorys* (Fig. 20.3, left), and there were other Upper Cretaceous echinoids (Table 20.1) and a crinoid columnal the size of a shirt button (Donovan, 2012). I even found a crinoid columnal in an erratic of Mississippian limestone (Donovan, 2010). But the diversity of Chalk echinoids has remained low.

Both are common fossils in the Upper Cretaceous, so why is *Echinocorys* so prevalent, and *Micraster* not, on this beach? I perceive there may be several possible explanations, and the answer is likely to be a combination of some and all of these, with or without other factors that I have failed to recognise. A list is the easiest way to visualise the various potential influences:

- In the Upper Cretaceous of north Norfolk, tests of *Echinocorys* may be much commoner than other echinoids. Potentially true, but proving it is problematic. Both *Echinocorys* and *Micraster* are common Chalk irregular echinoids. *Micraster* was an infaunal burrower. *Echinocorys* was epifaunal and robust and may be so common as to form well-recognised beds, such as at Margate (see Chapter 17).

- The big tests of *Echinocorys* are more obvious on the beach than other echinoids. Apart from *Echinocorys*, all the echinoids that I have found are somewhat smaller (Table 20.1) (compare in Fig. 20.3).

137

- The large tests of *Echinocorys* are particularly robust, both before final burial in the Late Cretaceous and after recent exhumation. *Echinocorys* tests formed 'benthic islands' following death on the Late Cretaceous seafloor. These dead tests shed their spines after death and were available for obligate encrusting invertebrates to attach. Once exhumed on the modern sea floor, the large tests of *Echinocorys*, filled with chalk or flint, are still robust and can also form modern benthic islands that are bored (Donovan, in press).

It is likely that the large size and robust tests of *Echinocorys* make it more obvious than *Micraster* on the north Norfolk coast. A further factor must be silicification. An echinoid in flint will be much more robust than one in chalk. *Micraster* is the only echinoid from this beach that has not been found in flint (Table 20.1), so far.

**Fossils in flint**: Further discussion of flint, so common on this beach, is desirable. Upper Cretaceous chalk is a white, particularly fine-grained limestone with relatively minor occurrences of flint (= chert), a secondary deposit commonly precipitated parallel to bedding and along faults or joints. The relative proportions of these rock types are reversed where they are reworked as beach clasts. Between Cromer and Overstrand, clasts are dominantly flint and chalk is a minor component of the beach load (Donovan, 2021b).

Numerous sponges and other fossils occur in rounded flint cobbles; their preservation may be either in cross-section or 'in the round'. Some flint groups, such as the sponges, are potentially identifiable in such sections, whereas other taxa are less easy to name (Ewin, 2018).

Herein, I discuss two contrasting fossiliferous flint clasts from the Norfolk coast. It is their unusual, yet contrasting, modes of preservation that make both of interest.

A sponge (NHMM 2020 028) is preserved as a well-rounded, elongate flint clast (Fig. 20.4). The sponge is incomplete but was possibly vase-shaped and probably oriented parallel to the long axis of the clast, whose shape it may have influenced during corrasion. It is slightly wider at the upper end in Figure 20.4, which exposes the irregularly shaped, grey-coloured fill of the atrium. In lateral view, the structure is a series of white and grey parallel bands.

## Table 20.1 Upper Cretaceous echinoids collected by the author from float between Overstrand and Cromer, north Norfolk coast, UK, since 2009

|  | Preserved in chalk | Preserved in flint |
|---|---|---|
| *Temnocidaris* (*Stereocidaris*) sp. ⸗ |  | + |
| phymosomatid sp. * |  | + |
| *Galerites* sp. | calcite test with a flint infill |  |
| *Micraster?* sp. * | + |  |
| *Echinocorys* ex. gr. *scutata* (Leske) | + | + |

Tests unless stated otherwise.
* = known from a single specimen.
⸗ = spine only.

**Figure 20.3** Upper Cretaceous *Micraster*? (right, NHMM 2021 010) and *Echinocorys scutata* ex. gr. (left, NHMM 2021 011), both in apical view (after Donovan, 2022a, fig. 4). From the beach between Overstrand and Cromer, north Norfolk. Specimens uncoated. Scale in cm and mm.

**Figure 20.4** Coiled shell encrusting a sponge (NHMM 2020 028) (after Donovan, 2022b, fig. 2). NHMM 2020 028, vase-like sponge preserved in flint (irregularly shaped atrium towards top) in lateral view, with an enigmatic coiled shell in intimate association in the lower half of the specimen. The coiled shell is most likely the serpulid *Proliserpula ampullacea* (J. de C. Sowerby). Specimen uncoated. Scale in mm and cm.

**Figure 20.5** Bored belemnite (NHMM 2020 029) (after Donovan, 2022b, fig. 3), *Belemnitella*? sp. (Left) NHMM 2020 030, partial belemnite rostrum preserved in calcite. (Right) NHMM 2020 029, external mould of partial belemnite rostrum preserving a dense infestation of borings, *Trypanites*? sp., in flint. Specimens uncoated. Scale in mm and cm.

The grey bands are thin and discontinuous; they most likely represent breaches in the wall of the sponge due to corrasion, exposing the fill of the atrium.

On one side of the sponge is a planispiral shell (Fig. 20.4), *c.*11.5 mm in diameter. Septa-like structures, although poorly seen, appear to be convex towards the aperture, but appear as radial grooves. These are probably not sections through septa, and are likely secondary, due to diagenesis or impacts after exhumation. The outer whorl is not in close contact with the inner whorls. The early whorls are replaced by an amorphous mass of flint. The aperture is oriented towards the atrium of the sponge.

The coiled shell is probably a serpulid worm. It seems most probable that it has been preserved because of its close association with the sponge (Fig. 20.4). Sponges have a spiculate endoskeleton that may be siliceous; this is why they are so common in flints. Being close to a sponge was doubtless a major part of the reason that an originally calcium carbonate conch was preserved in flint.

The close association of the specimens in NHMM 2020 028 gives a clue to the history of preservation. Most probably, this was a serpulid worm tube (= *Proliserpula ampullaceal* (J. de C. Sowerby); see Jäger, 2012) attached to the sponge in life. This identification favours an explanation for the close association between the sponge and the coiled shell, the two being cemented in close association in both life and death. Silicification occurred after final burial.

A second specimen is an incomplete belemnite rostrum preserved as an external mould, tapering towards a distal point (NHMM 2020 029; Fig. 20.5, right). The longitudinal section is off-centre. The rostrum is intensely bored close to the surface, with borings coarsely cast in flint. Borings are slender, straight to curved and apparently unbranched, but cross-cutting.

Belemnites in flint are commonly incomplete but are nonetheless easily identified as such. Although imperfect, the belemnite in Figure 20.5 (left) is most likely *Belemnitella* sp. The specimen preserved as an external mould (Fig. 20.5, right; NHMM 2020 029) is less obvious, but is notable for being densely bored. Bored belemnites are well known (such as Taylor *et al.*, 2013); infestation by encrusters and borers commonly occurred after the death of the belemnite. The righthand specimen is poorly preserved, but borings most likely represent *Trypanites*? isp. (compare with Taylor *et al.*, 2013, figs 15, 16).

**Taphonomy**: The examples discussed above demonstrate aspects of taphonomy, but here is one more specimen that shows a combination of features which, taken together, are worthy of close consideration. This specimen (Figs 2.5, 2.6) was immediately apparent to me on the beach because it included a fossil echinoid on the exposed side. *Echinocorys* is the commonest of erratic echinoids derived from the Chalk and known from the north Norfolk coast (see above). Thus, it was immediately intriguing to find this specimen, even before it was turned over (Donovan, 2023). See the detailed discussion in Chapter 2.

# References

Donovan, S.K. (2010) A Derbyshire screwstone (Mississippian) from the beach at Overstrand, Norfolk, eastern England. *Scripta Geologica, Special Issue*, 7: 43–52.

———. (2012) Taphonomy and significance of rare Chalk (Late Cretaceous) echinoderms preserved as beach clasts, north Norfolk, UK. *Proceedings of the Yorkshire Geological Society*, **59**: 109–113.

———. (2021a) *Hands-On Palaeontology: A Practical Manual*. Dunedin Academic Press, Edinburgh.

———. (2021b) Taphonomy of fossil invertebrates in flint beach clasts (Upper Cretaceous), north Norfolk coast. *Bulletin of the Geological Society of Norfolk*, **72**: 3–10.

———. (2022) A beachcomber's bonanza, or just another *Micraster*? *Geology Today*, **38**: 143–146.

———. (2023) Three views: Complex post-exhumation history of a Chalk cobble, north Norfolk. *Bulletin of the Geological Society of Norfolk*, **73**: 85–93.

———. (in press) Aspects of unusual taphonomy: reworked *Echinocorys* on Norfolk beaches. *Bulletin of the Geological Society of Norfolk*, **76**.

Donovan, S.K. & Lewis, D.N. (2011) Strange taphonomy: Late Cretaceous *Echinocorys* (Echinoidea) as a hard substrate in a modern shallow marine environment. *Swiss Journal of Palaeontology*, **130**: 43–51.

Ewin, T.A.M. (ed.) (2018) *British Mesozoic Fossils*. 8th ed. Natural History Museum, London.

Jäger, M. (2012) Sabellids and serpulids (Polychaeta sedentaria) from the type Maastrichtian, the Netherlands and Belgium. In Jagt, J.W M., Donovan, S.K. & Jagt-Yazykova, E.A. (eds) *Fossils of the type Maastrichtian (Part 1). Scripta Geologica, Special Issue*, **8**: 45–81.

Taylor, P.D., Barnbrook, J.A. & Sendino, C. (2013) Endolithic biota of belemnites from the Early Cretaceous Speeton Clay Formation of North Yorkshire, UK. *Proceedings of the Yorkshire Geological Society*, **59**: 227–245.

# CHAPTER 21

# Field guide: Easington, Co. Durham

## Preamble

> Industry has brought to the Durham coast, not only the unsightliness which seems to be its inevitable accompaniment, but also two alien and unpleasing deposits. […] at Easington waste material from the collieries has been pitched over the cliffs, despoiling the beach below and blackening the beaches to the south. (Ellis, 1954, p. 106)

This was true when Ellis's volume was published in 1954, but not today. Unfortunately, the so-called new edition (2018) of Ellis's fine book contains very few revisions, fails to make the transition to a plate tectonic paradigm and erroneously repeats the paragraph above. The colliery was closed by the Thatcher government in the 1980s and the beach is now 'a fabulous resource for pebble collecting' (Donovan, 2018, p. 798).

So, following the demise of the local coal mining industry, Easington is most worthy of our attention. Certain rock clasts, most particularly limestones and mudrocks, can be further broken down by the action of boring organisms which weaken them, particularly at the outer rim. This part of a rock clast may be densely infested by borers settling from single spatfalls. In contrast, invertebrates that form gregarious, cemented accumulations on mobile rock substrates may act as an additional layer of armour, retarding the breakdown of the clast. It is these borers and encrusters that make Easington such an intriguing site.

The beach north of Easington, Co. Durham (Fig. 21.1), retains many hundreds of rock clasts, including boulders, cobbles and pebbles. Most are Permian Magnesian Limestone derived from the cliffs at the back of the beach and from further north (by longshore drift), and from offshore, submerged outcrop; other lithologies were most likely derived from a mixture of longshore drift, reworked glacial erratics (Trechmann, 1931a) and bricks. There is ample evidence that the rock clasts have been rolled most energetically in the sea manifest by their common rounded shape; preservation on the beach would have been the work of storms in the North Sea. It is these rock clasts on the beach that are the subject of investigation, particularly those rich in invertebrate borings (Donovan *et al.*, 2018, 2019).

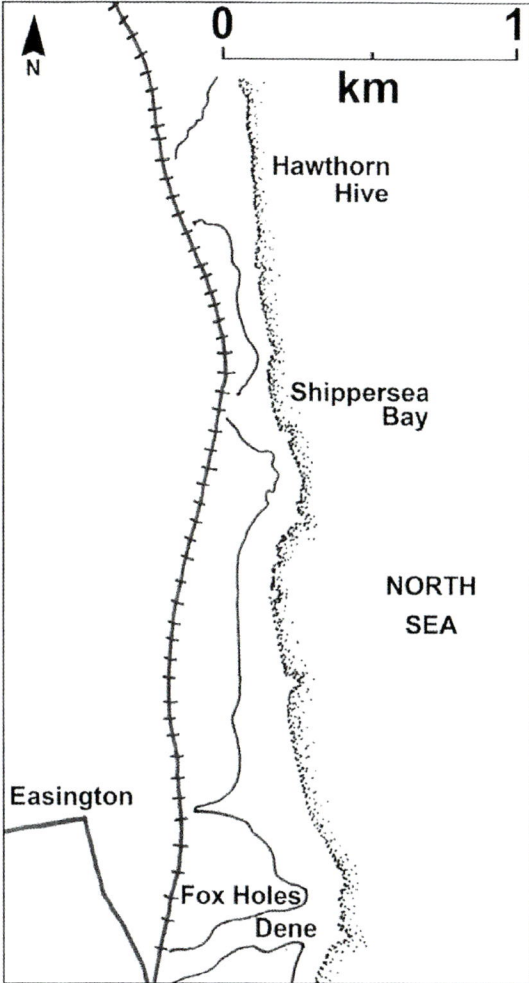

**Figure 21.1** Locality map of the coast north of Easington, Co. Durham (after Donovan *et al.*, 2018, fig. 1). Specimens described in this chapter were collected between Hawthorn Hive and Shippersea Bay. Key: heavy black line = major road; trellised line = railway; light black line = cliffline; stippled line = low water mark.

## How to get there

Travel to Durham by rail and take a bus or taxi to Easington (OS Explorer Sheet 308 'Durham & Sunderland'). If driving, find the car park on the B1283, east of the railway and Fox Holes Dean (Fig. 21.1; NGR NZ 439 435). Cross the railway and walk north between the railway and the cliff tops. Take the path down to beach level at Hawthorn Hive.

## Locality details

Easington's geology has a 'Jekyll and Hyde' relationship. The town is known for its colliery and the exploitation of coal from the Upper Carboniferous (Pennsylvanian). But this is hidden by the overlying marine Zechstein Group, Permian Magnesian Limestone, exposed particularly in the coastal cliffs of Co. Durham (Taylor *et al.*, 1971, pp. 70–76). The complexities of the Zechstein Group will not be untangled here, but see Ruffell *et al.* (2006, fig. 12.17 and supporting text). What is relevant to the current discussion is that the beach clasts are common, and dominantly derived from Permian limestones and dolostones, making them attractive substrates for modern boring and encrusting organisms.

## What to look for

This beach is rich in limestone (and other) clasts; shells are rarer but still occur and may be infested by other invertebrates. The specimens described below are a selection from samples seen between Hawthorn Hive south to below the Easington Raised Beach at Shippersea Bay (Fig. 21.1) and collected in July 2017 (Figs 21.2–21.7). Every attempt was made to sample the full range of boring and encrusting invertebrate taxa, and clast lithologies, seen on the day

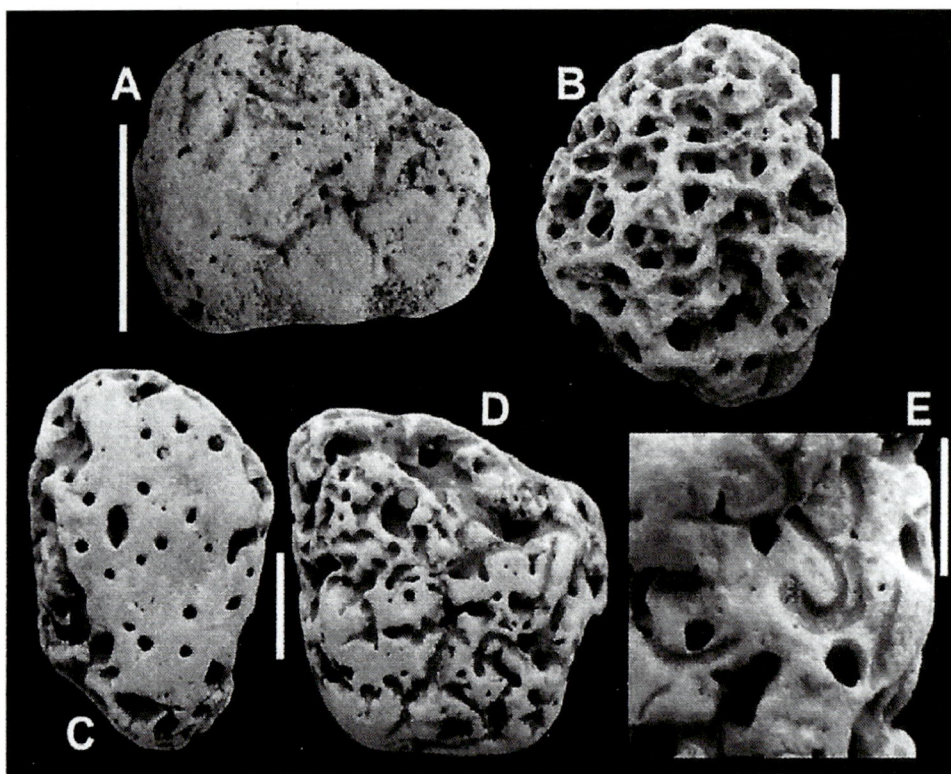

**Figure 21.2** (After Donovan *et al.*, 2018, fig. 2.) Magnesian Limestone cobbles bored by *Entobia* (**A–D**) and *Caulostrepsis*. (**A**) *Entobia* isp. aff. *E. cateniformis* Bromley & d'Alessandro, RGM 1332260, cobble that has been corraded to expose the internal network of borings, perhaps sub-parallel to the original surface. Scale bar represents 50 mm. (**B**) RGM 1332261, intensely bored pebble. (**C, D**) RGM 1332262, two views of a pebble, showing apertures on an external surface (**C**) and the internal colonial structures (**D**). (**E**) RGM 1332264, *Caulostrepsis* isp. aff. *C. spiralis* Pickerill *et al.* Scale bars represent 10 mm unless stated otherwise.

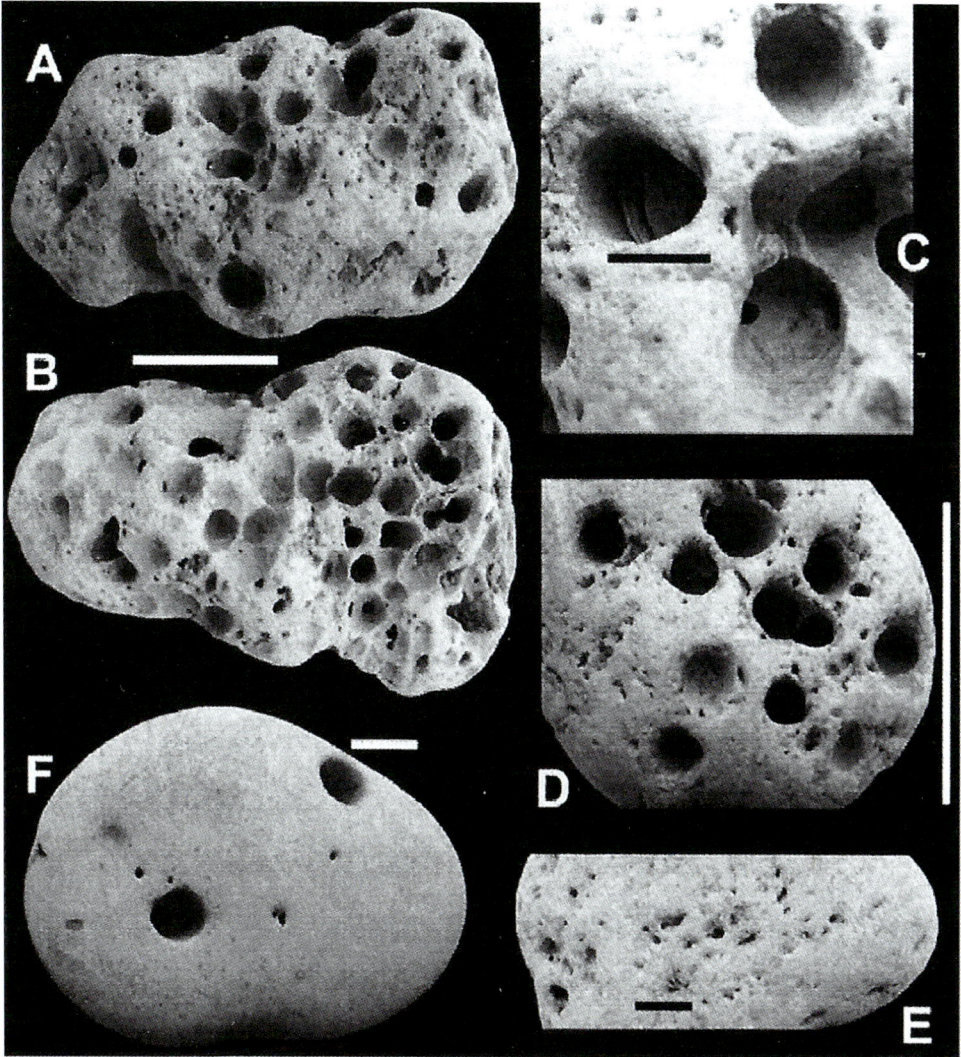

**Figure 21.3** (After Donovan *et al.*, 2018, fig. 3.) Magnesian Limestone cobbles bored by (mainly) *Gastrochaenolites clavatus* (Leymerie) (**A–D, F**) and *Caulostrepsis* isp. (**E**). (**A–C**) RGM 1332259. (**A, B**) Two sides of a cobble, showing incomplete borings of similar depth on both sides (*contra* specimen in Fig. 21.5), suggesting that this was a mobile clast bored equally on both sides and then similarly eroded correspondingly on both sides. Some borings have a calcite lining. Scale bar represents 50 mm. (**C**) Detail of boring, just above scale bar, with bivalve (borer? or nestler?) preserved within. It could not be removed without breaking and was therefore left *in situ*. (**D, E**) RGM 1332266. (**D**) Deep borings, *G. clavatus*, on one side of a cobble. Scale bar represents 50 mm. (**E**) Another side of the specimen showing slot-shaped borings (*Caulostrepsis* isp.). (**F**) RGM 1332263, unusually smooth limestone clast. The *G. clavatus* in the lower left is deep; that in the upper right is a hole through the clast and was bored from the reverse side. Scale bars represent 10 mm unless stated otherwise.

**Figure 21.4** (After Donovan *et al.*, 2018, fig. 4.) (**A, B**) RGM 1332269, cobble of Mississippian(?) limestone. (**A**) Densely bored surface of cobble. Scale bar represents 50 mm. (**B**) Detail of surface. Three good examples of *Caulostrepsis taeniola* Clarke are marked (\*). (**C**) RGM 1332270, limpet *Patella* sp. encrusted by basal attachments of serpulid *Pomatoceros triqueter* (Linné). The limpet is only encrusted in the area shown and not on the inner surface, which suggests that it may have been alive when infested. (**D**) RGM 1332271, gastropod *Nucella lapillus* (Linné) encrusted by scrpulid *Pomatoceros triqueter* (Linné) on the external surface only. (**E**) RGM 1332268, coal cobble encrusted by serpulid *Pomatoceros triqueter* (Linné), balanid *Balanus crenatus* Bruguière, calcareous algae *Lithothamnion* sp., spirorbids and bryozoans. The incomplete preservation of many encrusting organisms and the 'naked' areas of the clast indicates subsequent corrasion. Scale bar represents 50 mm. (**F, G**) RGM 1332267, sandstone cobble densely (**F**) to more sparsely infested (corraded) (**G**) by serpulid *Pomatoceros triqueter* (Linné), balanid *Balanus crenatus* Bruguière, calcareous algae *Lithothamnion* sp. and *Lomentaria*? sp., and bryozoans. Scale bar represents 50 mm. Scale bars represent 10 mm unless stated otherwise.

**Figure 21.5** (After Donovan *et al.*, 2018, fig. 5.) Two views of a limestone cobble that has been intensely bored by *Gastrochaenolites clavatus* (Leymerie) (specimen not collected). All of these borings would have been flask-shaped originally, so although side (**A**) appears to be more densely infested than (**B**), borings on the latter are shallower. One possible scenario would have been that side (**B**) was bored first, then partially corraded. Side (**A**) was then infested and corrasion continued equally on both sides, leaving those in (**A**) more complete. Scale bar represents 50 mm.

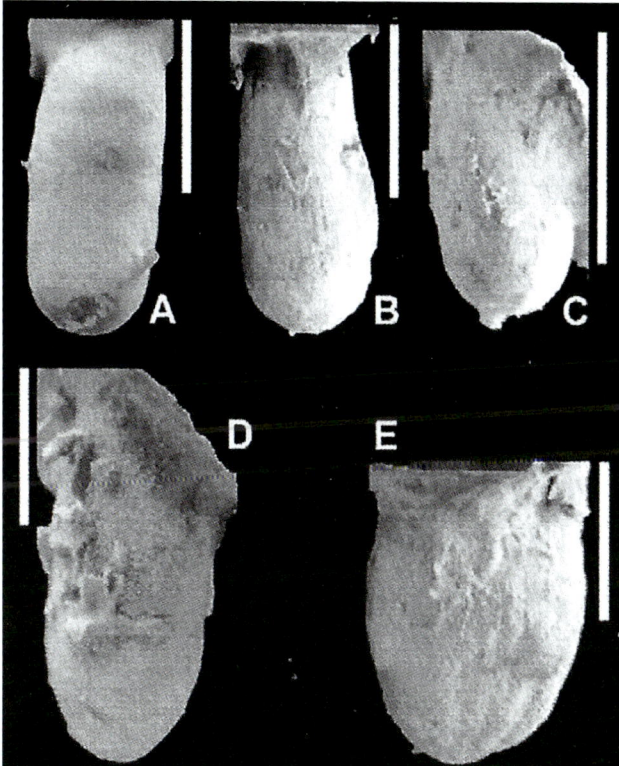

**Figure 21.6** (After Donovan *et al.*, 2018, fig. 6.) Latex casts of *Gastrochaenolites clavatus* (Leymerie). (**A**) RGM 1332263. (**B**, **C**) RGM 1332266, two specimens. (**D**, **E**) RGM 1332259, two specimens. All scale bars represent 10 mm. Specimens whitened with ammonium chloride.

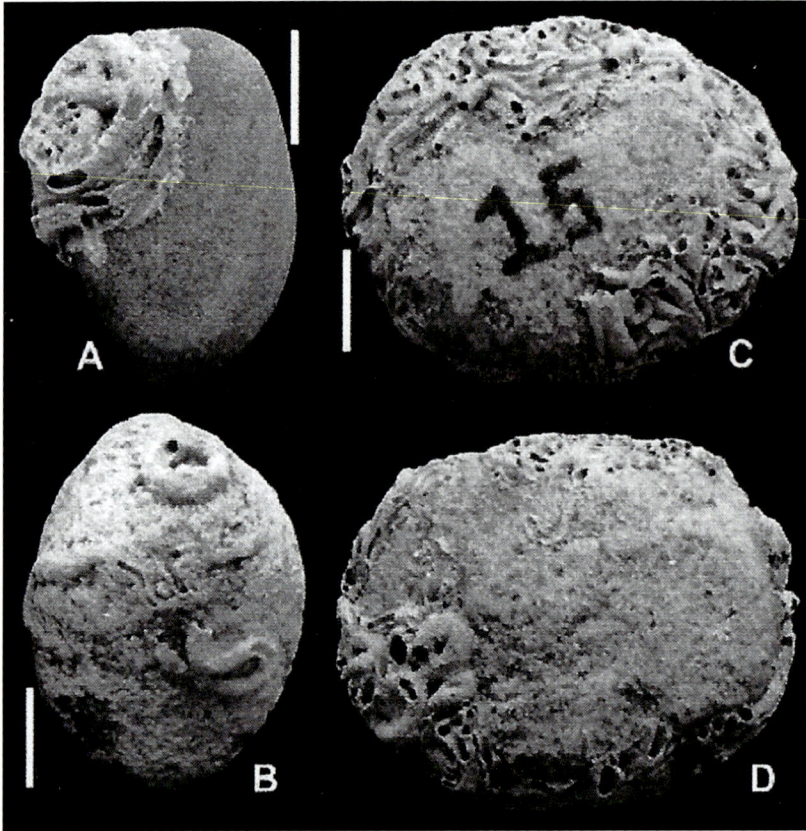

**Figure 21.7** (After Donovan *et al.*, 2018, fig. 7.) Encrusting organisms on sandstone pebbles. (**A**) RGM 1332272, pebble of fine-grained, bedded sandstone encrusted by serpulid *Pomatoceros triqueter* (Linné), which, in turn, preserves remnants of the calcareous alga *Lithothamnion* sp. This specimen is surely an erosional remnant of a clast that was formerly more densely covered by encrusters. (**B**) RGM 1332274, rounded pebble of coarse-grained sandstone, encrusted by serpulid worm tubes and subsequently overgrown by calcareous algae, *Lithothamnion* sp. (**C, D**) RGM 1332273, pebble of coarse-grained sandstone, densely encrusted in part by the serpulid *Pomatoceros triqueter* (Linné). The distribution of serpulids suggests that any tubes on the two flattened faces have been scraped clean during transport. All scale bars represent 10 mm.

they were collected. The variety of invertebrate encrusters on certain clasts is documented, providing a different pattern to that shown by borings. All collected specimens are deposited in the Naturalis Biodiversity Center, Leiden, the Netherlands (prefix RGM). This account is adapted from Donovan *et al.* (2018).

**Borings**: Three invertebrate ichnogenera are common at Easington and elsewhere on the North Sea coast (Donovan *et al.*, 2019). *Caulostrepsis* Clarke are 'U-shaped borings that have a vane connecting the limbs of the U-boring' (Bromley, 2004, p. 460) and are the spoor of polychaete worms, in the North Sea most commonly generated by the spionids that belong to the genus *Polydora*. They are typically shallow borings and are soon lost by surface

**Table 21.1 Summary of substrate preferences of trace fossils penetrating and invertebrates inhabiting pebbles and cobbles on the beach at Easington, based on a collection of 22 specimens (RGM 1332259 to 1332280; not all specimens are illustrated)**

| | | Miss Lst | Mag Lst | coal | sand-stone | gastro-pods | siltstone |
|---|---|---|---|---|---|---|---|
| **Borings** | | | | | | | |
| | *Caulostrepsis* | X | X | | | | |
| | *Entobia* | | X | | | | |
| | *Gastrochaenolites* | X | X | | | | |
| **Encrusters** | | | | | | | |
| | *Balanus* | | | X | X | | |
| | serpulids | 1 | 1 | X | X | X | X |
| | calcareous algae | | | X | X | | |
| | spirorbids | | | X | X | | |
| | bryozoans | | | X | X | | X |

Key: Miss Lst = Mississippian Limestone (Carboniferous); Mag Lst = Magnesian Limestone; 1 = encrusting unspecified limestone

corrasion of a mobile substrate. Where preserved, they are apparent in cross-section as slot or figure of eight-shaped holes (Fig. 21.3E) but are more apparent in longitudinal section. Straight specimens with a central vane, preserved in a cobble of Mississippian(?) limestone, are assigned herein to the type ichnospecies, *Caulostrepsis taeniola* Clarke (Fig. 21.4A, B). More teasing is a curved specimen, RGM 1332264, with an incomplete central vane and limbs that diverge more proximally (Fig. 21.2E). The curvature is reminiscent of *Caulostrepsis sprialis* Pickerill *et al.*, previously only recorded from the Middle Miocene of Carriacou, Lesser Antilles, but that ichnospecies lacks a central vane. It is provisionally referred to *Caulostrepsis* isp. aff. *C. spiralis*.

Sponge borings such as *Entobia* are 'generally an anastomosing network of canals that in most cases swell to form rounded chambers. Commonly the chambers dominate the boring and obscure the design of the network' (Bromley, 2004, p. 459). The interplay of ontogeny and taphonomy commonly makes identification of *Entobia* to ichnospecies testing except where specimens are particularly well preserved. Most specimens herein are assigned to *Entobia* isp. for simplicity (Fig. 21.2A–D). The complexities involved are demonstrated by RGM 1332262, which exposes both the internal (Fig. 21.2D) and external morphology (Fig. 21.2C), with apertures, of what is presumed to be a single network. RGM 1332260 is at least superficially close to *Entobia cateniformis* Bromley & d'Alessandro (Fig. 21.2A).

*Gastrochaenolites* includes clavate (club-shaped) borings in lithic substrates, including robust shells, like oysters, and wood (formerly included in *Teredolites* Leymerie, but now synonymised with *Gastrochaenolites*; Donovan & Ewin, 2018). These are the most prominent borings in rocks on the beach at Easington, partly because they are the largest borings in this assemblage, but also due to their high preservation potential. By boring vertical to sub-vertical to the surface of a clast, considerable corrasion is required to completely remove them (see, for example, Fig. 21.5), making *Gastrochaenolites* particularly persistent. The easiest way to determine the ichnospecific identity of modern *Gastrochaenolites* borings is to take casts of the borehole using some suitable medium, in this case liquid latex. All of the casts taken from RGM 1332259, 1332263 and 1332266 were similar (Fig. 21.6), although none was complete. Comparison with Kelly & Bromley (1984, text-fig. 3) shows that they are closest to *Gastrochaenolites turbinatus* Kelly & Bromley, a junior synonym of *Gastrochaenolites clavatus* (Leymerie) (Donovan & Ewin, 2018).

Apart from those infested by *Entobia* isp., the bored limestone clasts illustrated herein (Figs 21.3, 21.4A, B, 21.5) include indeterminate small, round or rounded holes that appear to be more or less deeply perforate in the substrate. Without seeing the three-dimensional form of these borings, it is impossible to determine to which ichnotaxa they should be assigned.

**Encrusters**: Gregarious accumulations of balanid barnacles, most likely *Balanus crenatus* Bruguière, occur on substrates of coal (Fig. 21.4E) and sandstone (Fig. 21.4F). On both specimens the balanids are partially overgrown by serpulid worms, implying a faunal succession. In turn, gregarious accumulations of small, presumed juvenile *B. crenatus* are found on some serpulids. On RGM 1332267 (Fig. 21.4F), the balanids and *Lithothamnion* show little evidence of interaction and may have been coeval.

Serpulid worms (Figs 21.4C–G, 21.7) are provisionally referred to *Pomatoceros triqueter* (Linné) (Campbell, 1982, pp. 134–135). They infest a range of substrates, sometimes densely, including gastropods, limestones (particularly within holes, such as vacant borings, particularly *Gastrochaenolites*), coal and sandstones, and have grown over *Lithothamnion* and *Balanus*, which in turn have encrusted *Pomatoceros*.

More than one specimen is encrusted by the white, chalky, warty calcareous alga referred to *Lithothamnion* sp. (Campbell, 1982, pp. 50–51; Figs 21.4E, F, 21.7B herein). It is possible that further algae are more common at Easington, but not apparent because of their similarity to the many pale limestone clasts that dominate the beach. RGM 1332267 preserves both encrusting *Lithothamnion* sp. and upright *Lomentaria*? sp. (Fig. 21.4F). RGM 1332274 is a sandstone pebble that is well-rounded, indicating considerable transport, and dense *Lithothamnion* sp. gives it the appearance of a limestone. It was encrusted by serpulid worms, and their tubes were subsequently overgrown by *Lithothamnion*. Some of the serpulid tubes have broken through, giving a false impression of sinuous borings (Fig. 21.7B, just above centre). Some algae have also been lost by corrasion. Rare, coiled spirorbid worm tubes, and incomplete (corraded) encrusting bryozoan colonies also occur.

The association of the boring 'trinity' of ichnogenera, *Caulostrepsis*, *Entobia* and *Gastrochaenolites*, is common on the coasts of the southern and western North Sea today (Donovan *et al.*, 2019), and elsewhere in the fossil record. These three ichnogenera can be easily separated by the novice ichnologist. The bored clasts are mobile and have been washed onshore, probably mainly during major storms, from the shallow shelf environment; they are all dwelling traces (= domichnia). Boring organisms only infest limestone substrates at this site; encrusters are recorded from various types of rock and shells, but not commonly on limestones.

It is instructive to compare the modern borings described herein with trace fossils from the Easington 60-foot raised beach (Oxygen Isotope Stage 7 = late Middle Pleistocene, *c.*38,000 years ago; Bridgland & Austin, 1999, p. 55; Davies *et al.*, 2009). Bridgland & Austin (1999, p. 53) included *Cliona* sp. and *Polydora* sp. in a faunal list and noted that 'Pebbles bored by marine molluscs and annelid worms are also common' (Woolacott, 1920, 1922; Trechmann, 1931b). *Cliona* sp. is unlikely, but this probably refers to clionaid sponge borings, *Entobia* isp. Similarly, the polychaete *Polydora* sp. would be a most unlikely fossil, but its borings, *Caulostrepsis* isp., are probable, particularly in *Ostrea* sp., which makes the same faunal list. Whether these are also the borings of 'marine […] annelid worms' is possible. The marine molluscs boring pebbles were most likely boring bivalves producing *Gastrochaenolites* isp. Woolacott (1920, pp. 308–310; 1922, p. 66) and Trechmann (1931b, p. 295) noted rolled clasts of Magnesian Limestone with boreholes inhabited by *Saxicava* Fleuriau de Bellevue, a junior synonym of *Hiatella* Bosc, a nestling and boring bivalve (Tebble, 1976, p. 173). Thus, available evidence suggests that the ichnotaxa of Easington have changed little since the Middle Pleistocene.

# References

Bridgland, D.R. & Austin, W.E.N. (1999) Shippersea Bay to Hawthorn Dene. In Bridgland, D.R., Horton, B.P. & Innes, J.B. (eds) *The Quaternary of North-East England: Field Guide*: 51–56. Quaternary Research Association, London.

Bromley, R.G. (2004) A stratigraphy of marine bioerosion. In McIlroy, D. (ed.) *The Application of Ichnology to Palaeoenvironmental and Stratigraphic Analysis*. Geological Society, London, Special Publication, **228**, 455–479.

Campbell, A.C. (1982) *The Country Life Guide to the Seashore and Shallow Seas of Britain and Europe*. 6th impr. Hamlyn, London.

Davies, B.J., Bridgland, D.R., Roberts, D.H., Cofaigh, C.O. *et al.* (2009) The age and stratigraphic context of the Easington Raised Beach, County Durham, UK. *Proceedings of the Geologists' Association*, **120**: 183–198.

Donovan, S.K. (2018) '"The Pebbles on the Beach: A Spotter's Guide" by Clarence Ellis, new edition'. [Book review.] *Proceedings of the Geologist's Association*, **129**: 798–799.

Donovan, S.K., Birtle, M., Harper, D.A.T. & Donovan, P.H. (2018) Borings and encrustations on cobbles and pebbles, Easington, Co. Durham. *Northumbrian Naturalist*, **85**: 49–61.

Donovan, S.K., with Donovan, P.H. & Donovan, M. (2019) (for 2018) A recurrent trinity of Recent borings in clasts around the southern and western North Sea. *Bulletin of the Geological Society of Norfolk*, **68**: 51–63.

Donovan, S.K. & Ewin, T.A.M. (2018) Substrate is a poor ichnotaxobase: A new demonstration. *Swiss Journal of Palaeontology*, **137**: 103–107.

Ellis, C. (2018) (first published 1954) *The Pebbles on the Beach*. Faber & Faber, London.

Kelly, S.R.A. & Bromley, R.G. (1984) Ichnological nomenclature of clavate borings. *Palaeontology*, **27**: 793–807.

Ruffell, A.H., Holliday, D.W. & Smith, D.B. (2006) Permian: arid basins and hypersaline seas. In Brenchley, P.J & Rawson, P.F. (eds) *The Geology of England and Wales*. 2nd ed. Geological Society, London, pp. 269–293.

Taylor, B.J., Burgess, I.C., Land, D.H., Mills, D.A.C. *et al.* (1971) *British Regional Geology: Northern England*. 4th ed. HMSO, London.

Tebble, N. (1976) *British Bivalve Seashells*. 2nd ed. HMSO, Edinburgh.

Trechmann, C.T. (1931a) The Scandinavian Drift or basement clay on the Durham coast. *Proceedings of the Geologists' Association*, **42**: 292–294.

———. (1931b) The 60-foot raised beach at Easington, Co. Durham. *Proceedings of the Geologists' Association*, **42**: 295–296.

Woolacott, D. (1920) On an exposure of sands and gravels containing marine shells at Easington, Co. Durham. *Geological Magazine*, **57**: 307–311.

———. (1922) On the 60ft raised beach at Easington, Co. Durham. *Geological Magazine*, **59**: 64–74.

# CHAPTER 22

# Field guide: Duart Bay, Isle of Mull, Inner Hebrides

## Preamble

Palaeontologists do not, in general, covet conglomerates. They may be fossiliferous, either in the matrix or, reworked, in the clasts (Donovan *et al.*, 2010; Donovan & Paul, 2013), but neither scenario is common. The high-energy environments in which conglomerates are deposited are not favourable to the preservation of shelly fossils, yet they do occur. One of the facets of *Aktuo-Paläontologie* that commonly has an advantage over the rock record is the hunt for shells in modern conglomeratic environments. For example, consider the preservation of shells, particularly molluscs, in modern conglomeratic environments, in this instance a beach rich in rock clasts on the south-east coast of the Isle of Mull (Figs 1.4, 22.1). This locality is dominated by lithic clasts; I have used it as a coarse model for the taphonomy and ichnology of some ancient conglomerates (Donovan, 2023).

The islands of the Inner Hebrides are largely volcanic. At first sight they may appear to be unlikely sites for the attentions of palaeontologists. The beaches are dominated by pebbles, cobbles and boulders of igneous origin. I have a strong interest in the distinctive taphonomic biases that such environments inflict on allochthonous shell assemblages, derived from the shallow offshore. This is an interest of the palaeontologist but does not necessarily run parallel with the studies of marine biologists. To put it another way, a palaeontologist on the seashore gathers information of relevance to the geologic past as much as the present. The Isle of Mull offers localities of relevance to understanding ancient conglomerates.

## How to get there

Most trips to the Isle of Mull 'start' at Oban. That is, the ferry from Oban to Craignure is the route chosen by most drivers and pedestrians to get to the Isle. The trip to the island takes less than an hour. Oban is reached by rail, and the station is adjacent to the ferry terminal. Other, minor ferries exist, linking smaller ports to the island, but they lack rail links. I recommend sailing from Oban.

On the Isle you will need your own transport, car or bicycle. Duart Bay is close to Craignure, and, on my last trip, I stayed with two friends in a cottage which was walking distance from Duart Bay. They also owned a car that enabled us to explore beaches further afield (Donovan, 2023).

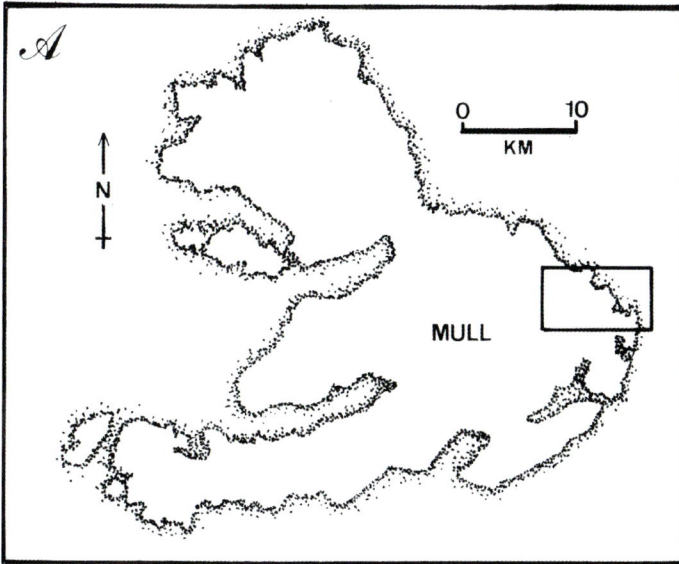

**Figure 22.1** Locality map. **A.** Outline map of the Isle of Mull; coastline stippled. Area of main map outlined (after Donovan, 2023, fig. 1). **B.** Outline map of study area, around Craignure (C) and Duart Castle (X) on Duart Point. Coastline stippled; roads in solid lines; dotted line shows route of the ferry from Oban to Craignure. Key: DB = Duart Bay; coarse stipple = shore at low tide in study area.

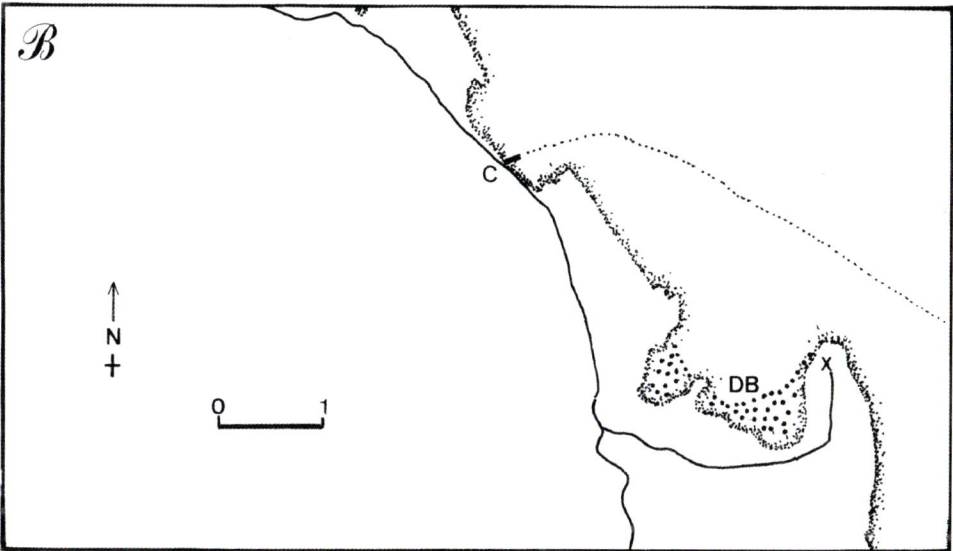

## Locality details

South and east sides of Duart Bay, south-east Isle of Mull, Inner Hebrides, western Scotland [about NGR NM 740 349 to 746 352] (Figs 1.4, 22.1). Collecting close to the beach is recommended, with access controlled in part by the tide. This is a pebbly, sandy beach with lithic (mainly igneous) clasts up to the size of boulders. Shells are not common and rarely well preserved, yet it is this poor preservation that provides an interesting taphonomic signal. For example, whelks are distinctive, with spires complete and shells broken away

**Figure 22.2** *Buccinium undatum* (Linné) (after Donovan, 2023, fig. 3). Eight whelks from Duart Bay (chosen from a collection of 30) showing the variation in preservation from the near-complete (upper right) to just the spire (lower left). Damage to shells was presumably mainly mechanical (see text). The only prominent encrusting organisms is *Balanus* sp. (bottom row, second from right). Specimens deposited in the Hunterian Museum, Glasgow. Scale bar in cm and mm.

(Fig. 22.2). Although limestone cobbles are rare, I collected two which were pitted by mature *Entobia* isp. (see below). In truth, few shells had borings of any kind that were identifiable in the field. Balanids and serpulids occurred prominently as encrusters on whelks and limpets; the latter were also common anchor points for wracks, which also provided sheltered points of attachment for balanids.

## What to look for

Adapted in part from Donovan (2023, pp. 2–5).

*The Boring Trinity*: As I have emphasised repeatedly, three shell- and rock-boring ichnogenera are common around the coast of the British Isles, namely *Caulostrepsis* Clarke, *Entobia* Bronn and *Gastrochaenolites* Leymerie (Donovan *et al.*, 2019). They are also common in shells. This does not imply identical producing organisms, but, rather, similar environmental responses within the *Trypanites* Ichnofacies. Note that the specimens referred to this ichnofacies herein are allochthonous, transported from the shallow water offshore by a range of natural phenomena.

*Caulostrepsis* and *Entobia* are common in Duart Bay, infesting shells. But where is *Gastrochaenolites*? Borings as big as a typical *Gastrochaenolites* need, at the least, cobbles of limestone or particularly large and robust shells, like oysters (such as Donovan *et al.*, 2014a), neither of which is common on the beaches of Mull. Although lithoclasts are common, limestone cobbles were rare at this locality. Only two bored cobbles of limestone were collected from Duart Bay and only preserved mature *Entobia* isp. While *Gastrochaenolites* may be rare even at sites where limestone clasts are moderately common, its absence at Duart Bay must be due, in part, to a paucity of suitable substrates.

**Shell breakage**: Whelks are common in Duart Bay as the spire and varying portions of the outer shell, more or less broken (Fig. 22.2). In part, this may have been the work of predatory crabs (e.g., Schäfer, 1972, fig. 237) or seagulls, but it is impossible to separate such breakage by predators from mechanical damage in the present study. However, the wave pounding received on the beach, with abrasion by rocky clasts, is probably the principal reason for damage. Similarly, it was impossible to say whether they had been weakened by borings (most likely *Entobia* isp.) before breakage proceeded to completion; infestation by borers may pre- or post-date mechanical damage.

The broken whelk shells (*Buccinum undatum* (Linné)) illustrated in Figure 22.2 are part of a collection of 30 such specimens that I made from this site. Although impossible to

**Figure 22.3** *Patella vulgata* Linné, Duart Bay (after Donovan, 2023, fig. 5). Three poorly preserved limpet shells seen in apical view, preserving *Caulostrepsis taeniola* Clarke and spirorbids (left); numerous spirorbids (centre); and *Balanus* with an attachment for wrack (right). Specimens deposited in the Hunterian Museum, Glasgow. Scale bar in cm and mm.

prove, as explained above, most of the damage was probably mechanical on a beach and foreshore rich in pebbles, cobbles and boulders. Damage is mainly external, to the shell, and not to the columella. Borings, notably the sponge spoor *Entobia* isp., are found on the shell, not the columella, suggesting weakening of the former by borings. Conversely, it is the inner and (mostly) outer surfaces of the shells that are encrusted by balanids, bryozoans and other taxa that would add to the strength (compare with Donovan *et al.*, 2014b). Overall, the shell is broken away first, leaving the robust columella, more internal and originally protected. There is undoubtedly a time factor involved; most specimens, although broken, have a 'fresh' shell.

***Encrusters***: Palaeontologists and biologists look at a beach in different ways. Palaeontologists have an interest in many aspects of 'life' that escape the biologist, such as neoichnology (see above) and taphonomy. Neoichnologic evidence is limited, but straightforward to interpret for both producers and functions. These data are supplemented by the evidence of calcareous encrusting invertebrates – *Balanus* (Fig. 22.2), serpulids and spirorbids (Fig. 22.3) – and attachments of wrack (sea weed) (Fig. 22.3).

The encrusting spirorbid worms are commonly well preserved (Fig. 22.3), yet are so small that, even when entire, incomplete (damaged) specimens may have been overlooked. It may be that at least some of these spirorbids may be late infestations that invaded the substrate shortly before they were collected. Serpulids are invariably incomplete, worm tubes probably broken by abrasion. Some *Balanus* are complete (Fig. 22.2), but many are identified from their basis plates only (Miller & Brown, 1979). What this 'sequence' (from poorly (serpulids) to better-preserved (spirorbids) skeletal encrusters) demonstrates is uncertain, but may indicate increasing shell strength, a biological succession or both.

***Discussion***: It is important to say that, although I have worked on fossiliferous conglomerates, the data presented herein are not directly applicable to my own experiences. I have hitherto worked on fossiliferous conglomerates in Jamaica, such as the late Pleistocene Farquhar's Beach red beds (Donovan *et al.*, 2010) and the Paleogene Richmond Formation (Trechmann, 1924; Donovan *et al.*, 1994). Both are notable for their well-preserved fossil gastropods, the former terrestrial and the latter marine. Contrast the variable preservation of whelks at Locality 1 herein (Fig. 22.2), with the well-preserved specimens illustrated by Trechmann (1924, pl. 2) and Donovan & Paul (2013, figs 3–5).

My strong impression of conglomerates, based on many years of observation, is that they tend towards two 'end members': either they preserve fossils quite well (such as Trechmann, 1924; Donovan & Paul, 2013) or not at all. It must be assumed that good preservation is produced by rapid burial in comparison with many examples (such as Brett, 1990; Fürsich, 1990; Anderson, 2001). If burial is slow or sporadic, broadly similar to a beach environment, then bioclasts have ample time to be broken up and ground down.

The preponderance of gastropods over bivalves at this locality is notable. The present collection includes 10 limpets, 30 whelks and 8 miscellaneous gastropods, but only 5 bivalves (including two that are articulated). This is counterintuitive. Many of the gastropods are predatory (e.g., the whelk *Buccinum undatum*; Beedham 1972, p. 56), yet they far outnumber the various prey species of bivalves. This is a nonsense – in life, predator species are invariably outnumbered by their prey. This is a strong indication of bias in the sample, due to hydrodynamic sorting, corrasion or a mixture of these and other factors.

## References

Anderson, L.C. (2001) Transport and spatial fidelity. In Briggs, D.E.G. & Crowther, P.R. (eds), *Palaeobiology II*. Blackwell Science, Oxford, 289–292.

Beedham, G.E. (1972) *Identification of the British Mollusca*. Hulton Educational Publications, Amersham.

Brett, C.E. (1990) Destructive taphonomic processes and skeletal durability. In Briggs, D.E.G. & Crowther, P.R. (eds), *Palaeobiology: A Synthesis*. Blackwell Scientific, Oxford, 223–226.

Donovan, S.K. (2023) Notes on *Aktuo-Paläontologie* of the rocky beaches of the eastern Isle of Mull, UK. *Scottish Journal of Geology*, **59**: 5 pp.

Donovan, S.K., Blissett, D.J. & Jackson, T.A. (2010) Reworked fossils, ichnology and palaeoecology: an example from the Neogene of Jamaica. *Lethaia*, **43**: 441–444.

Donovan, S.K., Cotton, L., Ende, Conrad van den, Scognamiglio, G. & Zittersteijn, M. (2014b) Taphonomic significance of a dense infestation of *Ensis americanus* (Binney) by *Balanus crenatus* Brugière, North Sea. *Palaios*, **28** (for 2013): 837–838.

Donovan, S.K., Dixon, H.L. & Veltkamp, C.J. (1994) Some echinoid plates from the Lower Eocene Richmond Formation of Jamaica. *Caribbean Journal of Science*, **30**: 145–148.

Donovan, S.K., with Donovan, P.H. & Donovan, M. (2019) (for 2018) A recurrent trinity of Recent borings in clasts around the southern and western North Sea. *Bulletin of the Geological Society of Norfolk*, **68**: 51–63.

Donovan, S.K., Harper, D.A.T., Portell, R.W. & Renema, W. (2014a) Neoichnology and implications for stratigraphy of reworked Upper Oligocene oysters, Antigua, West Indies. *Proceedings of the Geologists' Association*, **125**: 99–106.

Donovan, S.K. & Paul, C.R.C. (2013) Late Pleistocene land snails from 'red bed' deposits, Round Hill, south central Jamaica. *Alcheringa*, **37**: 273–284.

Fürsich, F.T. (1990) Fossil concentrations and life and death assemblages. In Briggs, D.E.G. & Crowther, P.R. (eds), *Palaeobiology: A Synthesis*. Blackwell Scientific, Oxford, 235–239.

Miller III, W. & Brown, N.A. (1979) The attachment scars of fossil balanids. *Journal of Paleontology*, **53**: 208–210.

Schäfer, W. (edited by G.Y. Craig) (1972) *Ecology and Palaeoecology of Marine Environments*. Oliver & Boyd, Edinburgh.

Trechmann, C.T. (1924) The Carbonaceous Shale or Richmond Formation of Jamaica. *Geological Magazine*, **61**: 2–19.

# Glossary

***Aktuo Paläontologie*** Examines live and dead organisms as fossils in the making. 'An actuopalaeontologist is naturally interested in almost every aspect of marine biology […]. However, he has certain additional interests that are normally not shared by his biological colleagues' (Schäfer, 1972, p. 2).

**allochthonous** Fossil organisms, shells or other clasts that have been transported. Such transport is commonly laterally, with shells being displaced by bottom currents or carcasses sinking to the bottom post-mortem, or vertically, with shells being displaced by burrowing organisms. See also autochthonous and parautochthonous.

**ammonite** An extinct group of nektic marine molluscs related to extant octopus and squid (the cephalopods).

**angular unconformity** Evidence of an interruption in sedimentation. Below the unconformity, beds were deposited, lithified, tilted by tectonic action and planed off by erosion. New sedimentation is horizontal, unlike the sloping beds below.

**annelid** Members of the Phylum Annelida are the segmented worms, such as the terrestrial earthworms in your garden. In the marine beach environment, shelly annelids are most commonly encountered as calcareous tubes, commonly secreted by groups (serpulids, spirorbids) that encrusted shells.

**aragonite** A polymorph of calcium carbonate, $CaCO_3$, chemically identical to calcite, but with different physical characteristics. Mineral in the shells of certain taxa, such as scleractinian corals and many gastropods.

**assemblage** In the context of this volume, a group of specimens (shells, fossils) found together (see death assemblage, life assemblage).

**autecology** The study of the ecology of individual organisms, especially functional morphology.

**autochthonous** Organisms preserved where they existed in life. This is not the same as a life assemblage, which preserves organisms that lived together although not necessarily in that place. For example, consider a Cretaceous nektic life assemblage – ammonites, fishes, mosasaurs – which may be preserved together on a bed even though they lived elsewhere, in the water column. See also allochthonous and parautochthonous.

**balanid** *see* **barnacle**

**barnacle** Multi-plated, gregarious encrusting organisms (Chapter 11). They are arthropods, that is, related to trilobites, crabs and insects. The unmineralised arthropod body of the barnacle was inside the shell and is commonly not preserved, but the shells are common fossils in Cenozoic deposits. Acorn barnacles (balanids) are conical and common on rocky shores. Goose barnacles have a stalk and are most likely found as pseudoplankton. Acrothoracian barnacles are borers, producing small, sock-shaped pits (*Rogerella* isp.).

**beachcombing** The pursuit of objects, exotic or otherwise, on the beach. In the area of study of the present volume, this is the search for shells, fossils and bored clasts, but other beachcombers may hunt, for example, archaeological relics.

**beachrock** Beachrock is produced predominantly in tropical, carbonate-rich, intertidal environments by penecontemporaneous cementation in the zone between high and low tides. Commonly composed of mainly carbonate grains with a carbonate (aragonite and/or high-magnesium calcite) cement, but non-carbonate beaches can be so lithified.

**benthic island** A robust dead shell on an unlithified substrate which provided potential surfaces of attachment for sessile, attached invertebrates such as barnacles and oysters.

**bioglyph** 'Bioglyphs are features in burrow or boring walls produced by such animal activity as scratching, drilling, plucking, gnawing, poking, and etching' (Ekdale & De Gibert, 2010, p. 540).

**biostratinomy** The study of that part of a fossil's history between death (including cause of death) and final burial. Essentially, this is an examination of the carcass (mainly the skeleton) as a sedimentary particle.

**boring** A hole made into or through a hard substrate, such as shell, rock, bone or wood. The mechanism of a borer may be chemical (dissolution), physical (with a hard grinding organ) or a combination of both.

**boulder** A clast with a diameter greater than 256 mm (Chapter 3).

**boulder clay** Old term for a conglomeratic glacial deposit containing a lithologically diverse (polymict), unsorted association of clasts (pebbles, cobbles, boulders).

**breccia** A sedimentary rock composed of angular clasts greater than 4 mm in diameter. The clasts are commonly of varying composition (= polymict). See Chapter 6.

**'butterfly'** A dead bivalved shell preserved with the two valves in close association due to the retention of the ligament after the muscles have rotted and the shell opened.

**calcite** A polymorph of calcium carbonate, $CaCO_3$, chemically identical to aragonite, but with different physical characteristics. Mineral in the shells of many taxa, such as barnacles and echinoderms.

**clast** Grains that make up siliciclastic sedimentary rocks. My own preference is to call fragments in sandstones and more fine-grained sedimentary rocks grains, and in more coarse-grained sedimentary rocks clasts, but they are, in truth, all clasts and all grains.

**clionaid sponge** The boring sponges which produce the trace *Entobia* ispp.

**cobble** A lithic clast with a maximum dimension 4–64 mm (see Chapter 3).

**colonial coral** Corals can be individual (= solitary) or colonial, the latter formed by the splitting into two or more branches.

**conglomerate** A sedimentary rock composed of rounded clasts greater than 4 mm in diameter (Chapter 3). The clasts are commonly of varying composition (= polymict). Fossils in clasts are derived from an older rock; fossils in the sedimentary rock between clasts are likely to be contemporary with the deposition of the rock.

**corrasion** A word compounded from corrosion + abrasion (Brett & Baird, 1986), used in describing a worn clast where the relative contributions of the two elements are not separable.

**death assemblage** A group of fossils in a bed and/or on a bedding plane that did not live together in life and became mixed post-mortem. For example, to give a real example, the Miocene Grand Bay Formation of Carriacou in the Lesser Antilles preserves a range of organisms from the terrestrial environment (land snails) to open ocean plankton, washed together by turbidity currents in deep water.

**dextral** Spiralling to the right as in gastropods coiling clockwise (e.g., Fig. 22.2).

**diagenesis** The changes, mainly chemical, that occur to a fossil after final burial. These can vary from very little, such as when a calcite (calcium carbonate) shell is buried in a limestone, to dissolution, recrystallisation and chemical replacement.

**disarticulation** The falling apart of multi-element skeletons, commonly post-mortem, but it can happen before death, such as leaves falling from a tree. Common organisms with complex multi-element skeletons that have a good fossil record include plants, vertebrates, echinoderms and arthropods.

**echinoid** A sea urchin, a member of the phylum Echinodermata with a globular, flattened or heart-shaped test (shell).

**encruster** An epifaunal organism that cements to a hard substrate.

**epifaunal** A general term describing those organisms that live or lived on the surface of the sediment or on the body of another organism. The former would include oysters and mussels; the latter includes acorn barnacles and serpulid worms.

**erratics** Clasts transported by physical action, such as by glaciers or as part of the bedload of rivers and accumulated far from their point of origin.

**functional morphology** The study of how a fossil worked as an organism.

**geomorphology** The study of the form of the Earth's surface.

**goose barnacle** *see* **barnacle**

**heart urchin** A heart-shaped echinoid (sea urchin). A typical example is the well-known Chalk *Micraster* (Fig. 20.3, right).

**ichnology** The study of traces and trace fossils, such as tracks, trails, burrows and borings.

**infaunal** A general term describing those organisms that live or lived beneath the sediment surface such as, for example, burrowing bivalve molluscs.

**life assemblage** A group of fossil organisms, preserved in close association, that reflects an original relationship that they had when alive.

**lithic fragments** Clasts composed of fragments of rocks that have been broken down by physical corrasion.

**longshore drift** The lateral transport of beach sediments by the action of waves impinging at an angle to the seashore.

**mudflat (or mud flat)** A muddy beach in a low energy setting.

**mudlarking** Beachcombing in rivers (particularly the Thames) where the sediment is dominantly muddy. Mudlarkers are mainly concerned with archaeology.

**nekrofauna** Dead, floating carcasses that are distributed by wave and current action.

**neoichnology** A division of ichnology that focuses on modern traces.

**organism–organism interactions** The synecology of two or more organisms that are intimately associated, such as balanid barnacles encrusting a live gastropod.

**palaeoautecology** The study of the palaeoecology of individual fossil organisms, particularly their functional morphology and its relation to their environment.

**palaeoecology** The ecology of fossil organisms, including their own habit and habitat, and the relationship to all aspects of the biological, chemical and physical environment.

**palaeosynecology** The study of communities (two or more individuals) of the past, and their relationships to each other and their environment.

**parautochthonous** Organisms preserved close to where they existed in life. Final burial probably followed death closely and after only minimal transport. For example, there is a common suspicion, which can often be demonstrated, that shells preserved on a shallow continental slope (such as eastern North America) will have undergone little transport, at most, before final burial (Donovan, 2002). See also allochthonous and autochthonous.

**pebble** A lithic clast 4–64 mm maximum dimension (Chapter 2).

**penecontemporaneous** 'formed during or shortly after the deposition of the containing rock stratum' (Lapidus, 1990, p. 397).

**photic zone** The shallow seafloor where sunlight can penetrate and support marine plants, particularly algae. Extends to 100–200 m depth in clear water.

**polymict** A beach deposit of pebbles, cobbles or boulders, or a conglomerate or breccia, where the identifiable clasts are derived from a diversity of source rocks.

**preservation** Of the many organisms that have lived on the Earth, only a sample is found in the fossil record. Preservation of a once-living organism in the rock record follows myriad chance pathways of biological, chemical and physical modification. This results in a fossil, derived from a dead organism and sufficiently alike its original form to be worthy of study, but nonetheless altered post-mortem.

**provenance** 'The place of origin, derivation, or earliest known history' (Brown, 1993, p. 2392) of a clast or shell.

**pseudoplankton** An encrusting or boring organism that infests, for example, a floating clast (pumice), wood, dead cephalopod shell or plastic bottle. The organism itself is not planktonic, but attachment to a floating substrate makes it so.

**pumice** A vesicular igneous rock that may float and, thus, may be infested by encrusting, pseudoplanktonic organisms (Fig. 7.1).

**raised beach** An ancient beach *sensu lato* that is elevated above modern sea level. Either the land has risen, or the level of the sea has fallen, or both.

**reef** A word with a plethora of definitions. For our purposes, a reef is an organic structure growing up from the seafloor, forming an ecological association of diverse sessile (e.g., corals) and vagile benthic organisms (e.g., gastropods) with nektonic groups such as cephalopods and fishes.

**reference collection** A collection made by an individual, university or museum to illustrate and define the organic remains of a particular beach, fossil group or stratigraphic interval.

**residence time** The time that a clast or dead shell remains on the seafloor before burial or discovery. The longer the residence, the more likely that the substrate will be encrusted or bored.

**serpulids** Annelid worms that secrete a calcareous tube (Chapter 15).

**siliciclastic rocks** Sedimentary rocks composed of grains derived from pre-existing rocks (sedimentary and/or igneous and/or metamorphic) that were rich in silicate minerals, particularly quartz (Chapter 3).

**sinistral** Spiralling to the left as in gastropods coiling anticlockwise, a rarer condition to dextral (clockwise) coiling.

**sorting** The physical action whereby clasts or shells may be moved in a particular direction to form an accretion. For example, disarticulated bivalves may have the left and right valves separated into different accumulations due to their contrasting hydrodynamic properties.

**spatfall** The settling of marine larvae of a particular generation. For example, balanids are gregarious and shells of a similar size on a substrate probably settled at the same time.

**spirorbids** Annelid worms that secrete a spiral calcareous tube (Chapter 15).

**succession (faunal)** As used herein, a substrate on which one encrusting organism is overgrown by a different encruster.

**synecology** The study of communities and their relationships, both to the environment and themselves, where a community is a natural grouping of two or more species that live in close association.

**taphonomy** The study of the preservation of fossils, including cause of death, post-mortem changes before burial and diagenetic changes following final burial.

**trace fossil** The two common types of fossils are body fossils, such as parts of the hard skeleton such as a shell, tooth or bone, and trace fossils. The latter are evidence of organic activity, called by some frozen ecology, and include such artefacts as tracks and trackways (footprints), trails, burrows, borings and coprolites. It is rare that a trace fossil and its producing organism are preserved in close association, but notable when it happens.

**transport** The movement of a clast or shell, most commonly under the influence of physical actions such as gravity, or wave or current action.

**univalve** A shell composed of a single component, such as a gastropod or nautiloid.

## References

Brett, C.E. & Baird, G.C. (1986) Comparative taphonomy: A key to paleoenvironmental interpretation based on fossil preservation. *Palaios*, **1**: 207–227.

Donovan, S.K. (2002) Island shelves, downslope transport and shell assemblages. *Lethaia*, **35**: 277.

Ekdale, A.A. & De Gibert, J.M. (2010) Paleoethologic significance of bioglyphs: Fingerprints of the subterraneans. *Palaios*, **25**: 540–545.

Lapidus, D.F. (1990) *Collins Dictionary of Geology.* Collins, London.

*The New Shorter Oxford English Dictionary.* (Brown, L., ed.) (1993) 2 vols. Clarendon Press, Oxford.

Schäfer, W. (G.Y. Craig, ed.) (1972) *Ecology and Palaeoecology of Marine Environments.* University of Chicago Press/Oliver & Boyd, Chicago and Edinburgh.

# Index